Interestingly, some relief from today's woes may come from ancient human practices. While current agri-food production models rely on abundant supplies of water, energy, and arable land and generate significant greenhouse gas emissions in addition to forest and biodiversity loss, past practices point toward more affordable and sustainable paths.

Different forms of insect farming and soilless crop farming, or hydroponics, have existed for centuries. In this report the authors make a persuasive case that frontier agriculture, particularly insect and hydroponic farming, can complement conventional agriculture. Both technologies reuse society's agricultural and organic industrial waste to produce nutritious food and animal feed without continuing to deplete the planet's land and water resources, thereby converting the world's wasteful linear food economy into a sustainable, circular food economy.

As the report shows, insect and hydroponic farming can create jobs, diversify livelihoods, improve nutrition, and provide many other benefits in African and fragile, conflict-affected countries. Together with other investments in climate-smart agriculture, such as trees on farms, alternate wetting and drying rice systems, conservation agriculture, and sustainable livestock, these technologies are part of a promising menu of solutions that can help countries move their land, food, water, and agriculture systems toward greater sustainability and reduced emissions. This is a key consideration as the World Bank renews its commitment to support countries' climate action plans.

This book is the World Bank's first attempt to look at insect and hydroponic farming as possible solutions to the world's climate and food and nutrition security crisis and may represent a new chapter in the organization's evolving efforts to help feed and sustain the planet. We hope the book will ignite further discussions and inspire concrete actions toward fully capturing the vast opportunities provided by insect and hydroponic farming as part of revamped, high-performing food systems that provide healthy and sustainable diets for all.

INSECT AND HYDROPONIC FARMING IN AFRICA

AGRICULTURE AND FOOD SERIES

A strong food and agriculture system is fundamental to economic growth, poverty reduction, environmental sustainability, and human health. The Agriculture and Food Series is intended to prompt public discussion and inform policies that will deliver higher incomes, reduce hunger, improve sustainability, and generate better health and nutrition from the food we grow and eat. It expands on the former Agriculture and Rural Development series by considering issues from farm to fork, in both rural and urban settings. Titles in this series undergo internal and external review under the management of the World Bank's Agriculture and Food Global Practice.

Titles in this series

Insect and Hydroponic Farming in Africa: The New Circular Food Economy (2021)

What's Cooking: Digital Transformation of the Agrifood System (2021)

The Safe Food Imperative: Accelerating Progress in Low- and Middle-Income Countries (2019)

The Land Governance Assessment Framework: Identifying and Monitoring Good Practice in the Land Sector (2011)

Rising Global Interest in Farmland: Can It Yield Sustainable and Equitable Benefits? (2011)

Gender and Governance in Rural Services: Insights from India, Ghana, and Ethiopia (2010)

Bioenergy Development: Issues and Impacts for Poverty and Natural Resource Management (2009)

Building Competitiveness in Africa's Agriculture: A Guide to Value Chain Concepts and Applications (2009)

Agribusiness and Innovation Systems in Africa (2009)

Agricultural Land Redistribution: Toward Greater Consensus (2009)

Organization and Performance of Cotton Sectors in Africa: Learning from Reform Experience (2009)

The Sunken Billions: The Economic Justification for Fisheries Reform (2009)

Gender in Agriculture Sourcebook (2008)

Sustainable Land Management Sourcebook (2008)

Forests Sourcebook: Practical Guidance for Sustaining Forests in Development Cooperation (2008)

Changing the Face of the Waters: The Promise and Challenge of Sustainable Aquaculture (2007)

Reforming Agricultural Trade for Developing Countries, Volume 2: Quantifying the Impact of Multilateral Trade Reform (2006)

Reforming Agricultural Trade for Developing Countries, Volume 1: Key Issues for a Pro-Development Outcome of the Doha Round (2006)

Enhancing Agricultural Innovation: How to Go Beyond the Strengthening of Research Systems (2006)

Sustainable Land Management: Challenges, Opportunities, and Trade-offs (2006)

Shaping the Future of Water for Agriculture: A Sourcebook for Investment in Agricultural Water Management (2005)

Agriculture Investment Sourcebook (2005)

Sustaining Forests: A Development Strategy (2004)

All books in the Agriculture and Food series are available free at https://openknowledge.worldbank.org/handle/10986/2151

INSECT AND HYDROPONIC FARMING IN AFRICA
THE NEW CIRCULAR FOOD ECONOMY

Dorte Verner, Nanna Roos, Afton Halloran,
Glenn Surabian, Edinaldo Tebaldi, Maximillian Ashwill,
Saleema Vellani, and Yasuo Konishi

1 2 3 4 24 23 22 21

ISBN (paper): 978-1-4648-1766-3
ISBN (electronic): 978-1-4648-1767-0
DOI: 10.1596/978-1-4648-1766-3

Cover photos: © Dorte Verner / World Bank. Used with the permission of Dorte Verner / World Bank. Further permission required for reuse.

Cover design: Melina Yingling, World Bank, from a concept by Eliot Cohen and Dorte Verner, World Bank, and based on a series design by Critical Stages, LLC.

Library of Congress Control Number: 2021918234.

CONTENTS

Foreword xvii

Acknowledgments xix

About the Authors xxiii

Executive Summary xxv

Abbreviations xxxv

1. Introduction 1
- Context of the Problem 2
- Solutions to the Problem 5
- Viability 9
- Road Map 11
- Methodology 12
- Notes 13
- References 14

2. Food Security Context 17
- Highlights 17
- Food Security and Nutrition in Africa 19
- Food Supply 25
- Economic Structure of the Agriculture Sector 32
- Population Change in FCV Countries 37
- Climate Change in FCV Countries 42
- Annex 2A 45

Notes 49
References 49

3. Understanding Insect Farming 53
Highlights 53
Context of Insect Farming in Africa 55
Types of Insects That Can Be Farmed 61
Roles in Insect Farming for Civil Society, Government, and the Private Sector 67
Insect Farming's Nutritional Benefits 79
Insect Farming's Social Benefits 83
Insect Farming's Environmental Benefits 84
Insect Farming's Economic Benefits 93
Annex 3A 99
Notes 104
References 104

4. Mainstreaming Insect Farming 115
Highlights 115
Edible Insect Supply Chains in African FCV-Affected States 116
Urban and Rural Insect Markets 118
Drivers of the Edible Insect Market 121
Edible Insect Production Systems 131
Modeling the Potential of BSF in Zimbabwe 144
Annex 4A. Profiles of Potential Benefits Derived from Black Soldier Fly in 10 African Countries 167
Notes 186
References 187

5. Understanding Hydroponics 193
Highlights 193
About Hydroponics 194
Types of Hydroponic Systems 197
Required Inputs 203
Outputs 207
Advantages over Soil Agriculture 212
Limitations 216
Notes 225
References 226

6. Ways Forward 231
Phase 1: Establishing and Piloting 234
Phase 2: Scaling 239
Note 240
References 240

Boxes

1.1 Benefits from Frontier Agriculture for Countries Affected by Fragility, Conflict, and Violence 8

1.2 Farm-Level and Country-Level Surveys 12

3.1 Insect Farming and the Sustainable Development Goals 85

4.1 Costs Associated with an Experimental Cricket Farming Activity in Kenya's Kakuma Refugee Camp 129

5.1 Hydroponic Pilot Project in Kenya's Kakuma Refugee Camp 196

5.2 Comparing Lettuce Yields, Water Usage, and Growing Seasons between Traditional Soil Farming and Two Hydroponic Techniques—the Wicking Bed and Nutrient Film Techniques—in West Bank and Gaza 206

Figures

ES.1 Linear versus Circular Economy for Food Production and Consumption xxviii

ES.2 Developing a Circular Food Economy xxix

ES.3 Comparative Advantage of Frontier Technology Relative to Conventional Farming When $R \leq R^*$ xxx

ES.4 Supply Chain Integration versus Costs over Time xxxi

1.1 Prevalence of Undernourishment in African Fragile, Conflict, and Violence Countries, 2015–30 2

1.2 Share of the Population with Insufficient Food Consumption in African FCV Countries 3

1.3 Linear versus Circular Economy for Food Production and Consumption 6

1.4 The Circular Food Economy and Its Benefits Using the Frontier Agricultural Technologies of Insect Farming and Hydroponic Crop Agriculture 9

1.5 Comparative Advantage of Frontier Technology Relative to Conventional Farming When $R \leq R^*$ 10

2.1 The Four Dimensions of Food Security 19

2.2 Food Insecurity's Negative Feedback Loop 21

2.3 Undernourishment Rates in FCV versus Non-FCV Countries in Africa, 2001–18 22

2.4 Undernourishment Is Pervasive and Increasing among FCV Countries 23

2.5 Prevalence of Anemia in Women of Reproductive Age (15–49 Years), 2016 24

2.6 Prevalence of Stunting, or a Height-for-Age More Than Two Standard Deviations below the International Median, among Children Younger Than Five Years 27

2.7 Food Supply in 13 African FCV Countries, 2018 29

2.8 Changes in Food Supply in 13 African FCV Countries, 2014–18 30

2.9 Average Protein Supply in 13 African FCV Countries, 2018 31

2.10 Change in Per Capita Food Production Variability in 18 African FCV Countries, 2000–16 32

2.11 Food Exports as a Percentage of Merchandise Exports 33

2.12 Average Annual Output per Worker in Agriculture, Forestry, and Fishery in 13 African FCV Countries, Various Years 36

2.13 Average Annual Population Growth in 19 African FCV Countries from 2000–04 to 2015–19 38

2.14 Population Living in Urban Areas in African FCV Countries, 2000 and 2019 39

2.15 Net Migration in African FCV Countries, 2002–17 40

2.16 Number of Refugees, by Country of Origin, 2015 and 2019 41

2.17 Number of Internally Displaced Persons in 14 African FCV Countries, 2019 41

2.18 Change in Average Temperature from 2000 to 2016 in African FCV Countries 43

2.19 Changes in Renewable Freshwater Resources from 2002 to 2017 in African FCV Countries 43

3.1 Number of Direct and Indirect Jobs Created in the Insect Food and Feed Industry in Europe 58

3.2 Korean Government Framework for the Insects-for-Food-and-Feed Industry 73

3.3 Results of a Life-Cycle Assessment of the Climate Impacts from Farming Crickets, Producing Broiler Chickens, and Optimizing Cricket Farms in Thailand 86

4.1 Zimbabwe's Wild Harvested Mopane Caterpillar Supply Chain 117

4.2 Democratic Republic of Congo's Wild Harvested Edible Insect Supply Chain 118

4.3 Papua New Guinea's Wild Harvested Edible Insect Supply Chain 118
4.4 Nontribal Social Arrangements of Wild Harvested Edible Insects in Zambia's Kazoka Village 121
4.5 Rough Representation of the Farmed Edible Insect Value and Supply Chains 122
4.6 Supply Chain Integration versus Costs over Time 128
4.7 Price Changes from Market Segmentation and Outsourcing Production to Small-Scale Insect Producers 130
4.8 Cricket Value Chain 135
4.9 BSF Value Chain 145
4.10 Maize Food Supply Chain and Annual Waste Stream, Zimbabwe 148
4.11 Sugarcane Food Supply Chain and Annual Waste Stream, Zimbabwe 150
4.12 Soybean Food Supply Chain and Annual Waste Stream, Zimbabwe 152
4.13 Groundnut Food Supply Chain and Annual Waste Stream, Zimbabwe 154
4.14 Wheat Food Supply Chain and Annual Waste Stream, Zimbabwe 156
5.1 Hydroponic Systems and How They Are Set Up 198
5.2 Aquaponics Cycle 202
5.3 Advantages and Disadvantages of Hydroponic Systems 203
5.4 Inputs into and Outcomes of Aquaponics and Hydroponics 204
B5.2.1 Lettuce Yield, Water Use, and Number of Growing Seasons per Year for Two Hydroponic Systems and the Traditional Soil Method in West Bank and Gaza 206
5.5 Hydroponic Space, Water Needs, and Yields for Producing Kale, Spinach, and Cowpeas 213
5.6 How Hydroponics Supports the World Bank Group's Four FCV Pillars 217
6.1 Developing a Circular Food Economy 232
6.2 Institutional and Regulatory Framework for Farmed Insects as Food and Feed 237

Maps

1.1 Countries in Which Insect Farming Data Were Collected for This Report 13

2.1 Stunting Rates in Low- and Middle-Income Countries, 2015 26

3.1 Diversity and Abundance of Edible Insects in Africa 56

Tables

2.1 Cereal Balance Sheet for 13 African FCV Countries with Available Data, 2018 28

2.2 Agricultural Employment as a Percentage of Total Employment in African FCV Countries, 2000–19 34

2.3 Value Added from Agriculture, Forestry, and Fishing as a Percentage of GDP in 19 African FCV Countries, 2000–19 35

2.4 Population Living below the National Poverty Line in 18 African FCV Countries, Various Years 37

2.5 Annual Freshwater Withdrawals for Agriculture as a Percentage of Total Freshwater Withdrawals, 2002 and 2017 44

2A.1 Undernourished People, 2000–18 45

2A.2 Children under Age Five Who Are Stunted 45

2A.3 Industry (Manufacturing and Construction) Value Added, 2000–19 46

2A.4 Literacy Rate 47

2A.5 Refugees, by Country of Origin, African FCV Countries, 2000–19 47

2A.6 Refugees, by Asylum Country, 2005–18 48

3.1 Willingness to Taste a Cricket Product among Nationalities in Kenya's Kakuma Refugee Camp, 2016 60

3.2 Most Commonly Farmed Insect Species 62

3.3 Growth Periods and Cycles of the Insect Species Observed in the Farm-Level Survey 65

3.4 Insect Species Farmed for Food and Feed in Africa as Identified in the Farm-Level Survey in 2019 66

3.5 Stakeholders in the Insects as Food and Feed Industry and Their Roles and Functions, 2019 67

3.6 Korean Government Areas of Investment for the Country's Insect Sector 74

3.7 Details of Large-Scale Insect Farming Companies, Based on Information Available in 2019 77

3.8 Fat and Protein in Various Edible Insect Species 80

3.9	Feed Conversion Rates of Various Insect and Livestock Species 88
3.10	Substrate Use on African Insect Farms 90
3.11	Nitrogen, Phosphorus, and Potassium Content of Chicken, Cricket, and Black Soldier Fly Larvae Biofertilizers 91
3.12	Annual Production for a Cricket Farm of Eight Pens with 8.5 Annual Growth Cycles in Thailand 94
3.13	Annual Revenue from Insect Farming in the Republic of Korea 94
3.14	Estimated Prices and Volumes of Selected Farmed Insect Species 96
3.15	Examples of By-products and Co-products Sold by African Insect Farms 97
3A.1	Insect Consumption in African Fragility-, Conflict-, and Violence-Affected States 99
4.1	Calendar for Crop and Wild Insect Harvesting in Zimbabwe and Description of When Consumption Occurs 119
4.2	Key Factors Associated with the Costs of Small-Scale and Commercial Insect Production Systems 123
B4.1.1	Items Supplied to Cricket Farmers in Kenya's Kakuma Refugee Camp 129
4.3	Inputs and Cost of Producing 1 Kilogram of Crickets in Kenya 130
4.4	Productivity of Different African Palm Weevil Farming Systems 140
4.5	Value of BSFL Converting Fecal Sludge into Different End Products in Three African Cities 141
4.6	Prices of Protein Sources for Animal Feed Available in Kenya 144
4.7	BSF-Related Conversion Factors 146
4.8	Recoverable Maize-Derived Outputs for Black Soldier Fly Substrates, Zimbabwe 148
4.9	BSFL, Meal, and Frass Production from Maize Output Substrates, Zimbabwe 149
4.10	Recoverable Sugarcane-Derived Outputs for Black Soldier Fly Substrates, Zimbabwe 150
4.11	BSFL, Meal, and Frass Production from Sugarcane Output Substrates, Zimbabwe 151
4.12	Recoverable Soybean-Derived Outputs for Black Soldier Fly Substrates, Zimbabwe 152

4.13 BSFL, Meal, and Frass Production from Soybean Output Substrates, Zimbabwe 152

4.14 Recoverable Groundnut-Derived Outputs for Black Soldier Fly Substrates, Zimbabwe 154

4.15 BSFL, Meal, and Frass Production from Groundnut Output Substrates, Zimbabwe 155

4.16 Recoverable Wheat-Derived Outputs for Black Soldier Fly Substrates, Zimbabwe 156

4.17 BSFL, Meal, and Frass Production from Wheat Output Substrates, Zimbabwe 157

4.18 Output Summary of BSFL, Meal, and Frass from Five Key Crops, Zimbabwe 159

4.19 Livestock Protein Demand, Zimbabwe 160

4.20 Total Dry and Crude Protein Converted from Five Key Crops, Zimbabwe 160

4.21 Approximate Retail Value of BSFL Protein, Zimbabwe 160

4.22 Black Soldier Fly Larvae Frass Production, by Crop, Zimbabwe 161

4.23 NPK Values of Common Organic Fertilizers 162

4.24 Employment Generation Estimates Associated with Black Soldier Fly Breeding, Zimbabwe 163

4.25 GHG Emissions Reduction and Energy Savings from Using BSFL Meal and Frass Instead of Traditional Livestock Feed Production and Synthetic Fertilizers, Zimbabwe 164

4.26 Crop Waste from Maize, Sugarcane, Groundnut, Soybean, and Wheat and the Amount of BSFL and Frass It Can Produce, Aggregates for All of Africa 165

4.27 Livestock Protein Demand, Aggregates for All of Africa 165

4.28 Employment Generation from Black Soldier Fly Production, for 10 and 30 Percent Conversion Rates, Aggregates for All of Africa 166

4.29 GHG Emissions Reduction and Energy Savings from Using BSFL Meal and Frass Instead of Other Meals and Organic Fertilizers, for 10 and 30 Percent Conversion Rates: Aggregates for All of Africa 166

4A.1 Countries Ranked by Value of Agriculture 167

4A.2 Crop Waste, BSFL, and Frass, Nigeria 168

4A.3 Livestock Protein Demand, Nigeria 168

4A.4 Employment Generation, Nigeria 169
4A.5 GHG Emissions Reduction and Energy Savings Realized through BSFL-Derived Meal and Frass, Nigeria 169
4A.6 Crop Waste, BSFL, and Frass, Kenya 170
4A.7 Livestock Protein Demand, Kenya 170
4A.8 Employment Generation, Kenya 171
4A.9 GHG Emissions Reduction and Energy Savings Realized through BSFL-Derived Meal and Frass, Kenya 171
4A.10 Crop Waste, BSFL, and Frass, Arab Republic of Egypt 172
4A.11 Livestock Protein Demand, Arab Republic of Egypt 172
4A.12 Employment Generation, Arab Republic of Egypt 173
4A.13 GHG Emissions Reduction and Energy Savings Realized through BSFL-Derived Meal and Frass, Arab Republic of Egypt 173
4A.14 Crop Waste, BSFL, and Frass, Ethiopia 174
4A.15 Livestock Protein Demand, Ethiopia 174
4A.16 Employment Generation, Ethiopia 175
4A.17 GHG Emissions Reduction and Energy Savings Realized through BSFL-Derived Meal and Frass, Ethiopia 175
4A.18 Crop Waste, BSFL, and Frass, Algeria 175
4A.19 Livestock Protein Demand, Algeria 176
4A.20 Employment Generation, Algeria 176
4A.21 GHG Emissions Reduction and Energy Savings Realized through BSFL-Derived Meal and Frass, Algeria 176
4A.22 Crop Waste, BSFL, and Frass, Tanzania 177
4A.23 Livestock Protein Demand, Tanzania 177
4A.24 Employment Generation, Tanzania 178
4A.25 GHG Emissions Reduction and Energy Savings Realized through BSFL-Derived Meal and Frass, Tanzania 178
4A.26 Crop Waste, BSFL, and Frass, Morocco 179
4A.27 Livestock Protein Demand, Morocco 179
4A.28 Employment Generation, Morocco 180

4A.29 GHG Emissions Reduction and Energy Savings Realized through BSFL-Derived Meal and Frass, Morocco 180

4A.30 Crop Waste, BSFL, and Frass, Sudan 181

4A.31 Livestock Protein Demand, Sudan 181

4A.32 Employment Generation, Sudan 182

4A.33 GHG Emissions Reduction and Energy Savings Realized through BSFL-Derived Meal and Frass, Sudan 182

4A.34 Crop Waste, BSFL, and Frass, Ghana 183

4A.35 Livestock Protein Demand, Ghana 183

4A.36 Employment Generation, Ghana 184

4A.37 GHG Emissions Reduction and Energy Savings Realized through BSFL-Derived Meal and Frass, Ghana 184

4A.38 Crop Waste, BSFL, and Frass, Angola 184

4A.39 Livestock Protein Demand, Angola 185

4A.40 Employment Generation, Angola 185

4A.41 GHG Emissions Reduction and Energy Savings Realized through BSFL-Derived Meal and Frass, Angola 186

5.1 Examples of Human Food or Animal Feed from Hydroponic Crops 208

5.2 Financial Breakdown of the Hydroponic Green Fodder Production System in West Bank and Gaza 211

5.3 Variable Costs of a 1,035 Square Meter Hydroponic Cucumber Production Operation in Turkey 219

5.4 Total Costs of a 1,035 Square Meter Hydroponic Cucumber Production Operation in Turkey 220

5.5 Initial Investment Costs for Constructing a 23 x 45 Meter Greenhouse in Turkey 221

5.6 Net Financial Returns Obtained from a 1,035 Square Meter Hydroponic Cucumber Production Operation in Turkey 223

5.7 Profitability of Two Hydroponic Systems in West Bank and Gaza 223

5.8 Time, Harvest, and Input Costs for Three Hydroponic Systems and Comparison with Soil-Based Production 224

5.9 Net Present Value and Benefit-Cost Ratio for Three Hydroponic Systems Compared with Soil-Based Production and Purchasing Vegetables with a Five-Year Horizon and 1 and 10 Percent Discount Rates 225

FOREWORD

The mission of the World Bank is to end poverty and improve equity through shared prosperity. But hard-fought development gains are now under threat. Every day there are signs of deepening climate change, dwindling natural resources, and intensifying food and nutrition insecurity, amid a global pandemic that has challenged people's ability to afford a healthy diet.

Interestingly, some relief from today's woes may come from ancient human practices. While current agri-food production models rely on abundant supplies of water, energy, and arable land and generate significant greenhouse gas emissions in addition to forest and biodiversity loss, past practices point toward more affordable and sustainable paths.

Different forms of insect farming and soilless crop farming, or hydroponics, have existed for centuries. In this report the authors make a persuasive case that frontier agriculture, particularly insect and hydroponic farming, can complement conventional agriculture. Both technologies reuse society's agricultural and organic industrial waste to produce nutritious food and animal feed without continuing to deplete the planet's land and water resources, thereby converting the world's wasteful linear food economy into a sustainable, circular food economy.

As the report shows, insect and hydroponic farming can create jobs, diversify livelihoods, improve nutrition, and provide many other benefits in African and fragile, conflict-affected countries. Together with other investments in climate-smart agriculture, such as trees on farms, alternate wetting and drying rice systems, conservation agriculture, and sustainable livestock, these technologies are part of a promising menu of solutions that can help countries move their land, food, water, and agriculture systems toward greater sustainability and reduced emissions. This is a key consideration as the World Bank renews its commitment to support countries' climate action plans.

This book is the World Bank's first attempt to look at insect and hydroponic farming as possible solutions to the world's climate and food and nutrition security crises and may represent a new chapter in the organization's evolving efforts to help feed and sustain the planet. I hope the book will ignite further discussions and inspire concrete actions toward fully capturing the vast opportunities provided by insect and hydroponic farming as part of revamped, high-performing food systems that provide healthy and sustainable diets for all.

Juergen Voegele
Vice President, Sustainable Development
World Bank Group

ACKNOWLEDGMENTS

This project was developed and led by Dorte Verner (Lead Economist in the Agriculture and Food Global Practice, Africa Region, World Bank). The book's other authors included Nanna Roos (Professor, University of Copenhagen), Afton Halloran (Consultant), Glenn Surabian (Consultant), Edinaldo Tebaldi (Professor, Bryant University), Maximillian Ashwill (Consultant), Saleema Vellani (Consultant), and Yasuo Konishi (Global Development Solutions, LLC).

The team is grateful for valuable support from Holger Kray (Practice Manager) and appreciates guidance and peer review comments from World Bank peer reviewers Richard Damania (Chief Economist), Geeta Sethi (Advisor), Svetlana Edmeades (Senior Agriculture Economist), Gerry Charlier (Lead Agriculture Specialist), and Erick Fernandes (Lead Agriculture Specialist) and external peer reviewer Jeffery Tomberlin (Professor, Texas A&M University). The team thanks the project's advisors for inputs and suggestions, including Arnold van Huis (Professor, Tropical Entomology, Wageningen University), Kwanho Park (Rural Development Administration, Republic of Korea), and Jeffery Tomberlin (Professor, Texas A&M University). The team is also grateful for contributions from the Rural Development Administration of the Republic of Korea by Tae-Woong Hur (Administrator), Nam Sunghee (Director), Bang Hae-Seon (Director), and researchers Hwang Jae-Sam, Park Kwan-Ho, Kim Mi-Ae, and Yoon Hyung-Ju. The team acknowledges support from the International Technology Cooperation Center at the Rural Development Administration, including Suntay Choi (Director), Ho-sun Lee, and Yiseul Kim. The team acknowledges guidance from Robert Musundire (Professor,

ABOUT THE AUTHORS

Maximillian Ashwill is an international development expert and strategist with 20 years of experience working with the World Bank, Peace Corps, and US Agency for International Development. He is also a professional writer and editor with more than 50 credited publications. Maximillian holds degrees from the University of Wisconsin in Madison and The New School in New York City.

Afton Halloran, PhD, holds a postdoctorate position in the Department of Nutrition, Exercise and Sports at the University of Copenhagen and is an independent consultant in sustainable food systems transitions and a transdisciplinary scientist. She works with such organizations as the World Bank and the Nordic Council of Ministers. She has written books and research papers on food-related topics.

Yasuo Konishi was the Managing Director of Global Development Solutions, LLC (GDS), located in Reston, Virginia. Mr. Konishi had 34 years of experience in enhancing competitiveness, covering myriad sectors, with the work generally culminating in development and investment strategies for client donor organizations and multinational corporations. Mr. Konishi additionally had 28 years of experience as an international agricultural economist and industry development specialist. He was the architect of the GDS integrated value chain analysis tool and executed more than 150 value chain analyses in more than 45 countries around the world.

Nanna Roos, PhD, is an associate professor of human nutrition in the Department of Nutrition, Exercise and Sports at the University of Copenhagen. She has 20 years of research experience with a focus on food-based approaches

have engaged in both activities for hundreds of years. However, farming insects and hydroponic crops to achieve development goals is a novel and innovative development approach, especially for increasing climate resilience in vulnerable communities, including refugees or others, who live in resource-constrained areas that do not support conventional farming, such as cities and arid environments.

Within a year, African insect farming can generate crude protein worth up to US$2.6 billion and biofertilizers worth up to US$19.4 billion. That is enough protein meal to meet up to 14 percent of the crude protein needed to rear all the pigs, goats, fish, and chickens in Africa, according to the report's modeling of the annual production of black soldier fly larvae (BSFL) in Africa. The report estimates that through black soldier fly farming, the continent could replace 60 million tons of traditional feed production with BSFL annually, leading to 200 million tons of recycled crop waste, 60 million tons of organic fertilizer production, and 15 million jobs, while saving 86 million tons of carbon dioxide equivalent emissions, which is the equivalent of removing 18 million vehicles from the roads.

> *Insect farming serves as a secondary income generation activity for me. Consulting on insect farming alone gives me almost half of my annual salary as a government employee.*
>
> *—Ghanaian farmer*

THE PROBLEM

The lack of effective demand undermines food and nutrition security in Africa and FCV countries. Acute food insecurity is on the rise in Sub-Saharan Africa, where about one in five people is undernourished (FAO et al. 2019). The situation is worse in African FCV countries,[3] where 29 percent of the population experiences insufficient food consumption, compared with 18 percent of the population in Sub-Saharan Africa overall. Individuals cannot afford enough nutritious food to live a healthy life. Africa's population growth, climate change, and land degradation also complicate the region's long-term food and nutrition security. With the Sub-Saharan population projected to reach about 2.2 billion by 2050 (Suzuki 2019), food access and production needs are increasing significantly. Meanwhile, the African continent is one of the world regions most affected by climate change, which further limits its ability to improve food and nutrition security through conventional farming systems (FAO and ECA 2018).

The world's natural resource base does not have the carrying capacity to sustain the world's current agri-food model, especially for animal feed. The agri-food system accounts for about 30 percent of the world's total energy consumption (FAO 2016), and agriculture, particularly animal rearing, is the most energy-intensive phase (Jasinski et al. 1999). Moreover, the agri-food system acquires 80 percent of its energy needs from fossil fuels (Monforti-Ferrario and Pinedo Pascua 2015), agriculture accounts for 70 percent of the world's freshwater withdrawals (UNESCO 2016), and the shares are even higher in the nine African FCV countries with available data (World Bank 2020). This all demonstrates that the world's natural environment will not be able to sustain the global food production model's current levels of natural resource extraction.

> *A benefit of insect farming is that it's a source of employment and income for both young and old.*
>
> —*Kenyan farmer*

THE SOLUTION

Frontier agricultural technologies can turn a linear food economy into a circular food economy. Two examples of frontier agriculture are insect farming and hydroponic crop farming. Both technologies fit within a circular economic model and are natural alternatives to the prevailing food system. They are suitable for the unique context of Africa and FCV countries. Neither requires access to abundant land, natural resources, or wealth to implement, all of which are limiting factors for Africans. The world, especially African FCV countries, needs an integrated food production system that can feed everyone, everywhere, every day with nutritious food while providing economic benefits and protecting the environment. A circular food economy is meant to be such a system. It is a systemic approach to food production that is mutually beneficial to businesses, society, and the environment.[4] It is designed to produce food while eliminating waste and pollution. The linear food economy follows a "take-make-waste" model in which resources are exploited to produce food while generating waste and polluting the environment. By contrast, a circular food economy is regenerative by design while generating socioeconomic benefits. It gradually decouples food production from the consumption of finite resources by reintegrating waste, instead of new resources, as an input into the food production system (see figure ES.1). Moreover, small-scale operations can be established economically and do not require much labor per kilogram of production.

The top of figure ES.2 demonstrates how frontier agriculture fits within the circular food economy and generates benefits. To summarize, farmed insects feed on certain organic waste. This turns a liability of the linear food economy into a benefit or asset of the circular food economy, by reducing and reusing society's organic waste. The figure shows that farmed insects and hydroponic crops are sources of proteins, micronutrients, various oils, and biofertilizer. These insect protein sources are then converted into feed for fish and livestock, which humans consume. Humans can also consume farmed insects and hydroponic produce directly. The wastes from insect and hydroponic farming are then fed back into the system and used as substrate and organic fertilizer.

There are already hydroponic farms and nearly 1,000 insect farms in Africa, and each year the number of new entrants and markets around the world

FIGURE ES.3 Comparative Advantage of Frontier Technology Relative to Conventional Farming When $R \leq R^*$

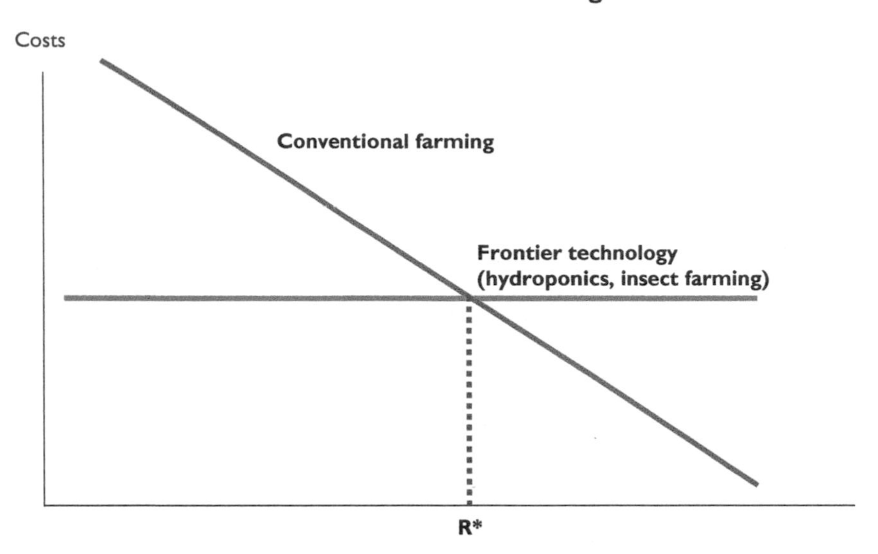

Source: Original figure for this publication.

VIABILITY

Frontier agricultural technologies have a cost advantage over traditional agriculture when resource availability is constrained. In certain situations, frontier agriculture is already more cost-effective than traditional agriculture. Figure ES.3 shows stylized cost curves for frontier farming and conventional farming. It shows that frontier agricultural technologies' cost advantages accrue in situations with tight resource constraints for water, arable land, and hard currency reserves. This is why these technologies have been successful in arid or densely populated areas, including in refugee camps, where there is a shortage of land and water. This also explains why frontier agricultural technologies are already economically advantageous alternatives to traditional agriculture in the resource-poor communities that are prevalent in Africa, including in FCV countries.

Frontier agriculture becomes even more cost-effective as these technologies are scaled up and processes become more specialized. In an integrated, or nonspecialized, production system, an individual producer carries out all the steps in the production process. In a specialized system, specialists carry out the steps separately according to their relative skills. As such, the costs per unit are lower in a specialized system than in an integrated system (figure ES.4). As these production systems reach scale, they are more likely to be specialized systems. Conversely, small-scale artisanal operations are more likely to be integrated and have higher marginal costs. These costs are largely determined

FIGURE ES.4 Supply Chain Integration versus Costs over Time

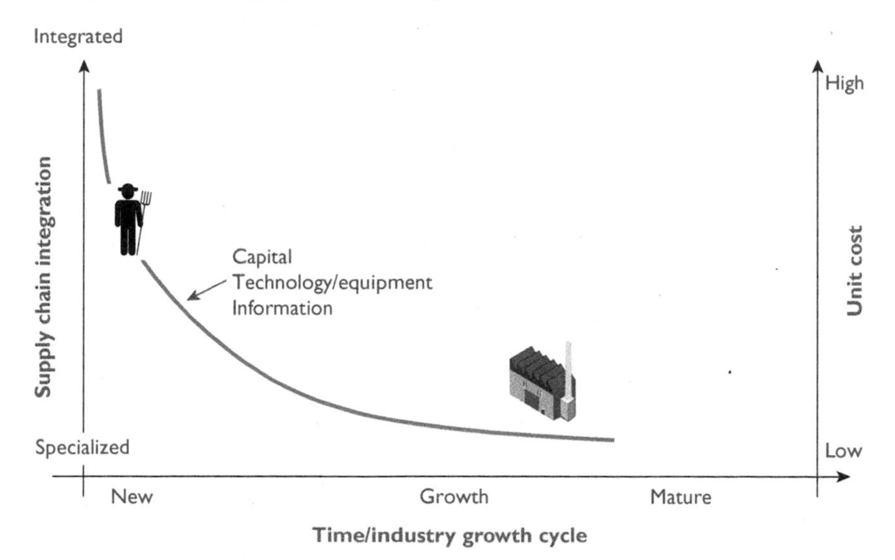

Source: Original figure for this publication.

by the system's access to capital, information, and technology and equipment. Moreover, as frontier agriculture becomes more prevalent and systems begin to scale, there will also be a greater likelihood of technological advancements, process innovations, and investments in research and development that will bring down costs, lowering the horizontal frontier cost curve (figure ES.3). Frontier technologies offer realistic prospects of scalability given conventional agriculture's large and growing demands on arable land, water, and energy resources.

> *The mopane caterpillar's manure fertilizes soil. The plants that use it are regenerated.*
>
> —*Congolese farmer*

WAYS FORWARD

The process of implementing a circular food economy based on frontier agricultural technologies can be organized into two phases. The first phase is to establish and pilot frontier agricultural systems, namely insect and hydroponic farming. To establish the necessary foundation of institutions and frameworks to carry the effort forward requires several key actions, such as training farmers, forming producer groups, and building producer capacity; providing access to finance; forming entomophagy and hydroponic associations; raising public awareness of the social, economic, and environmental benefits of hydroponic and insect farming agriculture; strengthening regulatory frameworks; and monitoring and evaluation. To pilot frontier agricultural systems would generate learning that could demonstrate their benefits and limit their inefficiencies. Such pilots would also improve the functionality of frontier agricultural operations and improve the operations' cost-effectiveness. The second phase is to scale up the frontier agricultural production systems at large enough levels to shift existing linear food economies into circular food economies. This will eventually bring down costs and enhance the competitiveness of insect farming and hydroponic agriculture. It will also reduce waste and protect the environment. Both phases will require action from the private and public sectors, including through public-private partnerships. For example, the public sector could provide extension services, necessary policies, or a regulatory framework, while the private sector could contribute starter capital or other investments. These two phases would also address the major factors that constrain the widespread adoption of insect and hydroponic farming in Africa, such as the general lack of knowledge, finance, and strong regulatory frameworks, among others. The bottom of figure ES.2 shows how the two phases—(1) establishing and piloting and (2) scaling—propel the circular food economy, leading to many direct socioeconomic benefits.

NOTES

1. Data from the World Food Programme, January 24, 2021.
2. FAOSTAT.
3. Calculation based on data from the World Food Programme, January 24, 2021.
4. This definition is adapted from that of the Ellen Macarthur Foundation (https://www.ellenmacarthurfoundation.org/explore/the-circular-economy-in-detail).

REFERENCES

AgriProtein. 2018. "USD 105 Million Raised for Sustainable Feed Firm: AgriProtein Secures Largest Investment to Date in Insect Protein Sector." Press Release, AgriProtein Online, June 4, 2018. https://www.agriprotein.com/press-articles/usd-105-million-raise-for-sustainable-feed-firm/.

Buhler Group. 2019. "Bühler Insect Technology Solutions and Alfa Laval Join Forces in Insect Processing." Buhler Group, Uzwil, Switzerland. https://www.buhlergroup.com/content/buhlergroup/global/en/media/media-releases/buehler_insect_technology solutionsandalfalavaljoinforcesininsect.html.

Byrne, Jane. 2018a. "Cargill Sees an Expansion of Its Functional Fish Feed Portfolio Globally, BioMar Evaluating Novel Proteins." *Feed Navigator,* March 13 2018. https://www.feednavigator.com/Article/2018/03/13/Cargill-sees-an-expansion-of-its-functional-fish-feed-portfolio-globally-BioMar-evaluating-novel-proteins.

Byrne, Jane 2018b. "McDonald's Championing Research into Insect Feed for Chickens." Reports from Feed Protein Vision 2018, *Feed Navigator*, March 27, 2018. https://www.feednavigator.com/Article/2018/03/27/McDonald-s-championing-research-into-insect-feed-for-chickens.

FAO (Food and Agriculture Organization of the United Nations). 2016. *The State of Food and Agriculture 2016: Climate Change, Agriculture and Food Security.* Rome: FAO.

FAO and ECA (Food and Agriculture Organization of the United Nations and European Commission on Agriculture). 2018. *Regional Overview of Food Security and Nutrition: Addressing the Threat from Climate Variability and Extremes for Food Security and Nutrition*. Accra, Ghana: FAO.

FAO, IFAD, UNICEF, WFP, and WHO (Food and Agriculture Organization of the United Nations, International Fund for Agricultural Development, United Nations Children's Fund, World Food Programme, and World Health Organization). 2019. *The State of Food Security and Nutrition in the World 2019: Safeguarding against Economic Slowdowns and Downturns*. Rome: FAO. https://docs.wfp.org/api/documents/WFP-0000106760/download/?_ga=2.96481119.1160401988.1576434696-1905750771.1576434696.

Jasinski, S. M., D. A. Kramer, J. A. Ober, and J. P. Searls. 1999. "Fertilizers." In *Sustaining Global Food Supplies*. United States Geological Survey Fact Sheet FS-155-99. Washington, DC: United States Geological Survey.

Law, C. 2020. "Insect Farming: The Industry Set to Be Worth $8 Billion by 2030." *Hive Life*, October 8, 2020. https://hivelife.com/insect-farming/.

MAFRA (Ministry of Agriculture, Food and Rural Affairs). 2019. "Survey of the Current Status of Korean Insect Industry." MAFRA, Government of the Republic of Korea, Sejong City.

MarketWatch. 2019. "Industry Research: Global Insect Feed Market Insights." News release, MarketWatch, New York, November 5, 2019. https://www.marketwatch.com/press-release/insect-feed-market-2019-industry-price-trend-size-estimation-industry-outlook-business-growth-report-latest-research-business-analysis-and-forecast-2024-analysis-research-2019-11-05.

Monforti-Ferrario, F., and I. Pinedo Pascua, eds. 2015. *Energy Use in the EU Food Sector: State of Play and Opportunities for Improvement*. Luxembourg: EU Publications Office.

Reuters. 2020. "Astanor Raises $325 Million Fund to Invest in Agri-Food Tech Startups." Reuters, November 20, 2020. https://news.yahoo.com/astanor-raises-325-million-fund-070000716.html?guccounter=1.

Suzuki, Emi. 2019. "World's Population Will Continue to Grow and Will Reach Nearly 10 Billion by 2050." *Data Blog*, July 8, 2019. https://blogs.worldbank.org/opendata/worlds-population-will-continue-grow-and-will-reach-nearly-10-billion-2050.

UNESCO (United Nations Educational, Scientific and Cultural Organization). 2016. *Water and Jobs: The United Nations World Water Development Report 2016*. Paris: UNESCO.

Weetman, C. 2016. *A Circular Economy Handbook for Business and Supply Chains: Repair, Remake, Redesign, Rethink*. London: Kogan Page Publishers.

Welborn, K. 2021. "2021 Mid-Year AgTech Venture Capital Investment Round Up." *CropLife*, June 29, 2021. https://www.croplife.com/precision/2021-mid-year-agtech-venture-capital-investment-round-up/.

Wilbur-Ellis. 2018. "New COO and NSF Grant Fuel Beta Hatch Growth and Efficiency." Wilbur-Ellis Company, San Francisco. https://www.wilburellis.com/new-coo-and-nsf-grant-fuel-beta-hatch-growth-and-efficiency/.

World Bank. 2020. *World Development Indicators*. Washington, DC: World Bank.

ABBREVIATIONS

°C	degrees Celsius
ACFS	National Bureau of Agricultural Commodity and Food Standards (Thailand)
ACP-EU	African, Caribbean, and Pacific–European Union
ALV	African leafy vegetables
ASF	animal source foods
BSF	black soldier fly
BSFL	black soldier fly larvae
BSG	brewer spent grains
CAGR	compound annual growth rate
CO_2	carbon dioxide
CO_2-eq	CO_2-equivalent
COVID-19	coronavirus
ECA	European Commission on Agriculture
EFSA	European Food Safety Authority
EPA	Environmental Protection Agency (United States)
FAO	Food and Agriculture Organization of the United Nations
FCR	feed conversion rate
FCV	fragility, conflict, and violence
g	grams

GAP	Good Agricultural Practice
GDP	gross domestic product
GHG	greenhouse gas
GRID	green, resilient, and inclusive development
GWP	global warming potential
ha	hectares
IAEA	International Atomic Energy Agency
IDP	internally displaced person
IEA	International Energy Agency
IFAD	International Fund for Agricultural Development
IFIF	International Feed Industry Federation
IFPRI	International Food Policy Research Institute
IITA	International Institute of Tropical Agriculture
Int$	international dollars
IPIFF	International Platform of Insects for Food and Feed
K	potassium
K Sh	Kenya shilling
kcal	kilocalories
kg	kilograms
l	liters
m	meters
m^2	square meters
MAFRA	Ministry of Agriculture, Food and Rural Affairs (Republic of Korea)
MENA	Middle East and North Africa
MWh	megawatt hours
N	nitrogen
NFT	nutrient film technique
NGO	nongovernmental organization
NPK	nitrogen, phosphorus, and potassium
NPV	net present value
P	phosphorus
PV	photovoltaics
PVC	polyvinyl chloride
R&D	research and development
RDA	Rural Development Administration (Republic of Korea)

SDGs	Sustainable Development Goals (United Nations)
SFA	Singapore Food Agency
t	tons
TNO	Organisation for Applied Scientific Research
UNCTAD	United Nations Conference on Trade and Development
UNDP	United Nations Development Programme
UNECA	United Nations Economic Commission for Africa
UNESCO	United Nations Educational, Scientific, and Cultural Organization
UNHCR	United Nations High Commissioner for Refugees
UNICEF	United Nations Children's Fund
US$	United States dollar
WFP	World Food Programme
WHO	World Health Organization
WMO	World Meteorological Organization

CHAPTER ONE

Introduction

The purpose of this report is to increase knowledge on food insecurity in Africa and present innovative solutions to address the challenge of increasing food security. The report assesses the benefits of expanding frontier agricultural technologies within a circular food economy in Africa, with a particular focus on countries affected by fragility, conflict, and violence (FCV). Frontier agricultural technologies are approaches to agricultural production that sustainably expand the frontiers of current food production practices. This report provides knowledge and solutions for accelerating the global response to rising hunger and malnutrition in Africa and getting the second Sustainable Development Goal of Zero Hunger back on track. These solutions would contribute to transforming food systems for healthier people, healthier economies, and a healthier planet. For 5 million years, humans and their hominin ancestors have consumed foraged insects and plants (Lesnik 2018). Much more recently, humans have started farming insects and using hydroponics to grow plants for human food and animal feed. Building on this history, the report examines the feasibility of expanding two frontier agricultural technologies—insect farming and hydroponics—in Africa to reduce food insecurity and increase green, resilient, and inclusive development (Development Committee 2021).

The report shows that frontier agriculture is a viable complement to conventional agriculture, particularly in Africa and countries affected by FCV. The report finds that adopting frontier agricultural systems in Africa can bring food security and provide many health, socioeconomic, and environmental benefits. The report proposes applying a circular economy concept to frontier agricultural technologies. Doing so would reuse society's organic and industrial waste to produce nutritious foods for humans and animals, such as fish and livestock. Such a system can build the food system's environmental sustainability and resilience to climate change;

pandemics, such as the COVID-19 (coronavirus) pandemic; and other shocks. The report proposes using insect farming and hydroponics to increase access to protein and other vital nutrients to reduce food insecurity in FCV-affected African countries and beyond.[1]

CONTEXT OF THE PROBLEM

An estimated 121 million people, or 24 percent of Sub-Saharan Africa's population, consume less nutritious food than they need for a healthy life.[2] After several years of decline, and since even before the COVID-19 pandemic, the prevalence of acute food insecurity has been on the rise in Sub-Saharan Africa, where about one in five people is undernourished, or 250 million individuals (FAO 2019). Chronic undernutrition has stunted one in three children in Sub-Saharan Africa and is the leading cause of death for 45 percent of the world's children under age five years, totaling 3.1 million preventable child deaths per year (Black et al. 2013).

FIGURE 1.1 Prevalence of Undernourishment in African Fragile, Conflict, and Violence Countries, 2015–30

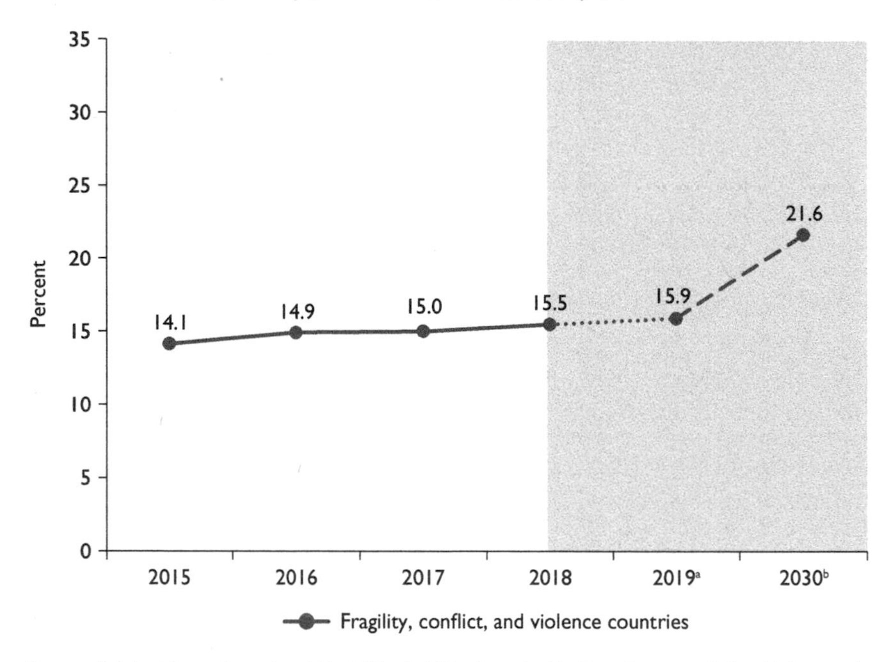

Source: Original figure based on World Bank Africa Sustainable Development Policy Unit compilation of data from FAO 2020.

Note: The graph underestimates the food insecurity situation as data from some of the most food-insecure fragility, conflict, and violence (FCV) countries, such as Niger and South Sudan, are not available and therefore not included.

a. Projected values.

b. Projections up to 2030 do not reflect the potential impact of the COVID-19 pandemic.

Undernourishment is even higher in FCV countries (FAO et al. 2017).[3] Figure 1.1 shows that undernourishment rates in African FCV countries will continue to rise for at least a decade, which will cause declines in childhood development, economic productivity, and people's general well-being, thereby leading to reduced economic development.

The food security situation is worse in African FCV countries, where 29 percent of the population experiences insufficient food consumption. This is higher than in Sub-Saharan Africa overall, where 18 percent of the population experiences insufficient food consumption. Figure 1.2 shows the share of the population that is food insecure in African FCV countries. In Burkina Faso, Mali, Niger, and South Sudan, more than half the population experiences insufficient food consumption. These impacts are particularly dire for already vulnerable populations, including refugees and internally displaced persons. Africa is home to 40 percent of the world's forcibly displaced persons (UNHCR 2020). Chapter 2 provides a detailed analysis of food insecurity in Africa and FCV countries.

The COVID-19 pandemic has pushed as many as 150 million people into extreme poverty globally. In Sub-Saharan Africa, the number of confirmed COVID-19 cases reached 5.8 million and the number of confirmed deaths surpassed 142,000 as of September 28, 2021.[4] Countrywide lockdowns due to

FIGURE 1.2 Share of the Population with Insufficient Food Consumption in African FCV Countries

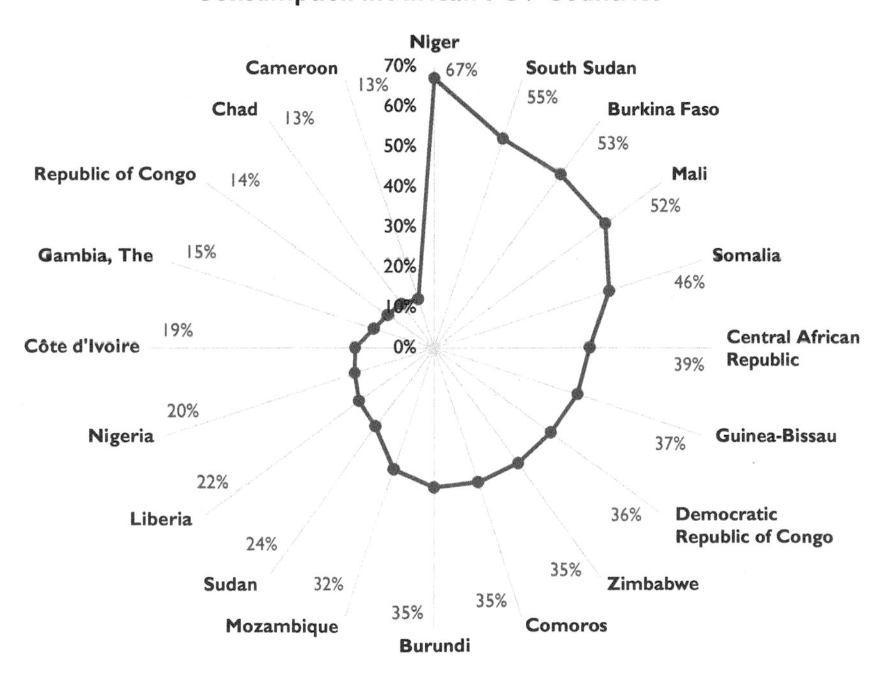

Source: Original figure based on World Food Programme data, January 24, 2021.
Note: FCV = fragility, conflict, and violence.

COVID-19 caused market disruptions and led to job and income losses. For example, in Nigeria, 42 percent of the respondents to a phone survey were no longer working, despite being employed pre-COVID-19. In Ethiopia, 45 percent of urban households and 55 percent of rural households reported income losses from COVID-19. Remittances from Africans working abroad have also dwindled. Like in most economies, COVID-19-related job losses have been larger for lower-income and informal workers whose jobs cannot be performed from home. These impacts are exacerbated by natural shocks, climate change, natural resource shortages, and agricultural challenges, such as crop and livestock pests and diseases, including the desert locust emergency (Dabalen and Paci 2020), in addition to lack of hard currencies for import of agriculture inputs and food, poor socioeconomic policies, and conflicts and fragility.

The rise in food insecurity has been one of the pandemic's most tangible symptoms, but the agriculture sector has been less affected. The Office of the World Bank's Chief Economist, Africa Region, estimates that the COVID-19 pandemic doubled acute food insecurity, decreased domestic food production by 7 percent, and reduced food imports by 13 to 25 percent for African countries, of which 39 are already net food importers. The World Food Programme estimated that by the end of 2020, more than 900 million people globally had insufficient food to eat, and projections suggest there will be an additional 9.3 million wasted children and 2.6 million stunted children in the world because of COVID-19. This will have negative impacts on the health and cognitive development of COVID-19-era children for years to come, which will lead to declines in human capital. That said, the share of African households involved in agriculture has been increasing since the start of the pandemic (Amankwah, Gourlay, and Zezza 2021). Overall, the agriculture sector seems to serve as a buffer for low-income households to fall back on when other income-generating activities are lost. Something similar happened during the 2008 global financial crisis. And when an economic crisis affected Thailand in the late 1990s, many rural workers returned to the northeast of the country and started to farm crickets for human food (see chapter 4).

Lack of effective demand drives the region's food insecurity. Effective demand is determined by a family's purchasing power, or its ability to pay for food and other goods and services. Therefore, the lack of demand is driven largely by income poverty and ineffective distribution. In 9 of the 20 African FCV countries, 50 percent of the population lives in poverty, with poverty levels reaching 82 percent in South Sudan (World Bank 2020a). Individuals living below the poverty line often cannot afford enough nutritious food to live a healthy life. In Africa, poverty and ineffective distribution are the largest contributors to food insecurity (World Hunger Education Service 2018).

Africa's population growth, land degradation, and climate variability and change also complicate the region's long-term food security. With the Sub-Saharan population projected to reach 2.2 billion people by 2050

(Suzuki 2019), food access and production needs are increasing significantly. Meanwhile, the African continent contributes relatively little to total global greenhouse gas (GHG) emissions, but it is one of the world regions most negatively affected by climate change, which further limits its ability to improve food security (FAO and ECA 2018). In Africa's FCV countries, increased climate variability is already causing more droughts, floods, and pest attacks, such as the 2020–21 locust emergency. Moreover, average temperatures have risen over the past two decades by more than half a degree Celsius in 17 of 19 African FCV countries.[5] Together with exploitative conventional farming techniques and declining freshwater resources, this has led to increased land degradation and, in some places, aridity in African FCV countries.[6] One study projects that climate change will reduce average crop yields in Africa by 2050, including reducing wheat yields by 17 percent, sorghum by 15 percent, millet by 10 percent, and maize by 5 percent (Knox et al. 2012). This is reinforced by undernourishment statistics, which show that the number of undernourished people in drought-sensitive countries has increased by 45.6 percent since 2012 (FAO et al. 2019).

The world's natural resource base does not have the carrying capacity to sustain the current agri-food model, especially for animal feed. The agri-food system accounts for about 30 percent of the world's total energy consumption (FAO 2016), with agriculture, particularly animal rearing, being the most energy-intensive phase (Jasinski et al. 1999). For example, soy production accounts for about 10 percent of Brazil's deforestation, with three-quarters of this soy being fed to livestock as animal feed.[7] Moreover, the agri-food system acquires 80 percent of its energy needs from fossil fuels (Monforti-Ferrario and Pinedo Pascua 2015). And agriculture accounts for 70 percent of the world's freshwater withdrawals (UNESCO 2016) and an even higher percent in the nine African FCV countries with available data (World Bank 2020a). Aquifers replenish so slowly that they are effectively a nonrenewable resource (Dalin et al. 2017). Studies also show that soil erosion rates under plowed cultivation are twice as fast as soil replenishment rates (Montgomery 2007). This creates a need for fertilizers, of which the vast majority still require fossil fuels to produce (Kudo and Miseki 2009). For example, the global agri-food system uses about 90 percent of the world's extracted phosphorus (Childers et al. 2011). These unsustainable land use practices also contribute to climate change and biodiversity loss, which exacerbate food supply challenges even more (World Bank 2020b). This all demonstrates that the world's natural environment will not be able to sustain the global food production model's current levels of natural resource extraction.

SOLUTIONS TO THE PROBLEM

Creating a "circular economy" model for food production would help solve Africa's food security challenges and make food production systems more sustainable. The world, especially African and FCV countries, needs an integrated

FIGURE 1.3 Linear versus Circular Economy for Food Production and Consumption

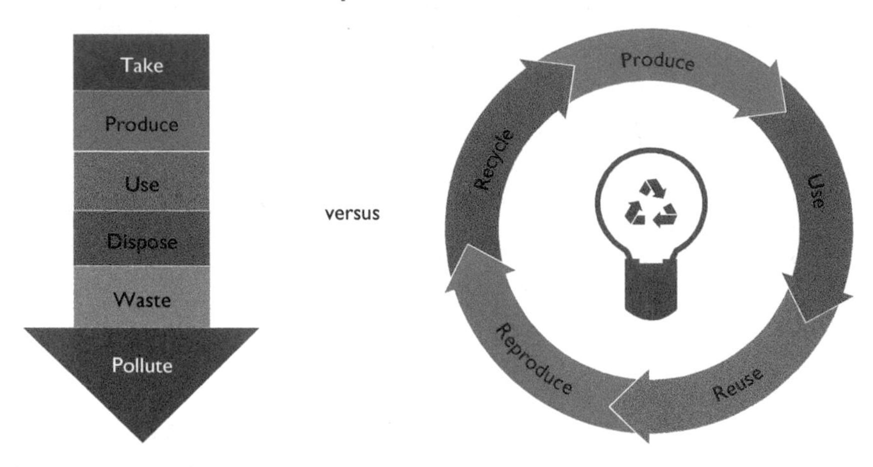

Source: Original figure for this publication, based on Weetman 2016.

food production system that can feed everyone, everywhere, every day with nutritious food while providing economic benefits and protecting the environment. A circular food economy is meant to be such a system. It is a systemic approach to food production that is mutually beneficial to businesses, society, and the environment.[8] It is designed to produce food while eliminating waste and pollution.

The circular economy and the traditional linear economy are shown in figure 1.3. The linear economy follows a "take-make-waste" model in which resources are exploited to produce food while generating waste and polluting the environment. For example, about 30 percent of the world's agriculture and food production is lost or wasted in the food supply chain. This model relies on continuous inputs of new resources to produce more outputs, and it has been the dominant model for economic development since the Industrial Revolution. By contrast, a circular economy is regenerative by design. It gradually decouples food production from the consumption of finite resources by reintegrating waste as an input into the food production system instead of new resources. With increasing demand for nutritious food and dwindling natural resources from climate change and environmental exploitation, a circular economy model is better suited to Africa's food security needs than the current linear economy. A circular food economy would allow the food system to produce food efficiently and sustainably while retaining value along the supply chain and preventing waste and pollution (waste-to-value) by keeping products and materials in use. Such a system would reduce food insecurity, create jobs and income, and regenerate natural systems.

Frontier agricultural technologies can turn a linear food economy into a circular food economy. Such technologies do not require arable land or significant water resources, contribute less to GHG emissions, and have a limited impact on the environment. Two examples of frontier agriculture are insect farming and hydroponic crop farming. Insect farming is the process of producing insects for human food and animal feed, and hydroponic farming is the process of growing crops in nutrient-rich water solutions instead of soil. Both technologies are alternatives to the prevailing food system and a natural fit for the unique context of African and FCV countries. Neither technology requires great access to land, space, natural resources, or wealth to implement, all of which are limiting factors for many Africans.

This report focuses on insect and hydroponic farming because they are less affected by the water and arable land constraints in African FCV countries. These constraints are especially harsh in poor and arid regions. Climate change is exacerbating aridity in Africa, making water scarcity even more of a resource constraint for poor farmers (WMO 2019). As this report shows, insect and hydroponic farming do not require large amounts of water or arable land and can make access to nutritious food more secure. This has important implications for the many areas of Africa affected by food insecurity and FCV (described in chapter 2). For example, evidence shows that just a 10 percent increase in the local food price index is associated with a 0.7 percentage point increase in violence against civilians (Gutiérrez-Romero 2020). There are other frontier agricultural technologies in addition to insect and hydroponic farming, but these other technologies are less conducive for African FCV countries. For example, the report does not consider algae farming despite its promising potential within a circular food economy because 12 of the 20 African FCV countries are landlocked or have only limited access to the sea and algae farming is currently most cost-effective in coastal areas.[9] The report also does not consider permaculture, which can be sustainable but requires arable land. This report's survey findings indicate that insect farming is becoming a fast-growing industry in Africa with many public and private sector stakeholders (see chapter 3) that have been working with these technologies for a few years. This means that local experts are available in Africa to guide new market entrants as the global market rapidly develops and expands.

Both insect farming and plant hydroponics offer benefits to African FCV countries. As the report shows, farmed insects can be a tasty source of animal protein and essential micronutrients for humans and animals alike. Insect farming can even reduce existing organic agricultural and industrial waste from the linear economy by using this waste as food for the farmed insects. Hydroponics is a source for essential fruits, vegetables, and fodder. These technologies can create economic opportunities in both rural and urban areas and provide livelihoods for vulnerable populations, including youth, women, and refugees. Frontier agricultural technologies also require less manual labor than conventional agriculture and livestock rearing, and small-scale operations can be established economically (see chapters 4 and 5). Both technologies can help food-importing countries become more food independent, or self-sufficient, by producing more food locally. The increased food self-sufficiency helps improve national accounts and can alleviate the stress on hard currency reserves faced by many African ministers of finance. Such reserves are often used for fertilizer or animal feed imports, such as soybeans from South America. The benefits of frontier agricultural technologies are listed in box 1.1.

Figure 1.4 demonstrates how frontier agriculture fits within the circular food economy and generates these benefits. To summarize, farmed insects feed on organic waste, including organic agricultural or food industrial waste. This turns a liability of the linear food economy into a benefit or asset of the circular food economy, by reducing and reusing society's organic waste. The figure shows that farmed insects and hydroponic crops are sources of protein, micronutrients, oil, and biofertilizer. These insect protein sources are then converted to feed for fish and livestock, which humans consume. Humans can also consume farmed insects and hydroponic products directly. The waste from insect and hydroponic farming is then fed back into the system and used as substrate and organic fertilizer.

BOX 1.1 Benefits from Frontier Agriculture for Countries Affected by Fragility, Conflict, and Violence

- Increased domestic production of nutritious foods and feed
- Reduced waste and pollution compared with the linear production model
- Improved sustainability of local food systems and natural resources because of less water requirements, reduced less land and biodiversity degradation, and fewer greenhouse gas emissions during the food and feed production process compared with traditional agriculture
- Improved soil health through application of organic fertilizers consisting of the insect manure (frass) produced during the insect farming process
- Improved macroeconomic situations and increased national savings of hard currency through reduction of domestic reliance on protein imports
- Increased access to jobs, incomes, and livelihoods, particularly along the food value chain
- Improved peacebuilding and resilience to fragility, conflict, and violence through the creation of more stable and sustainable food supply chains that provide economic opportunities and require fewer natural resources

FIGURE 1.4 The Circular Food Economy and Its Benefits Using the Frontier Agricultural Technologies of Insect Farming and Hydroponic Crop Agriculture

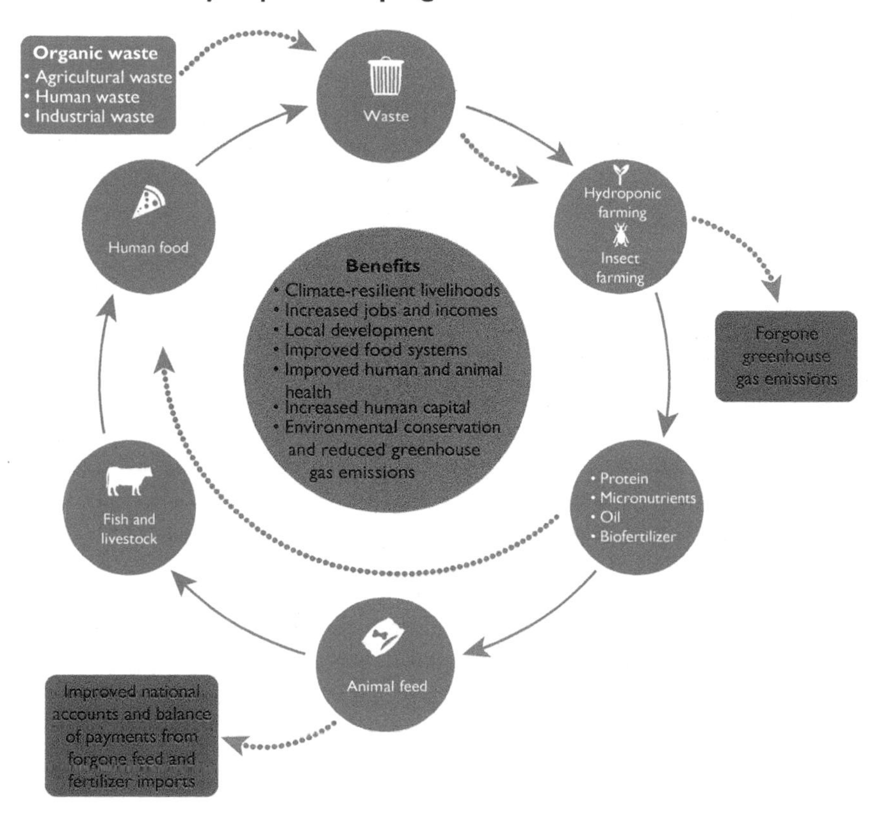

Source: Original figure for this publication.

VIABILITY

Frontier agricultural technologies have a cost advantage over conventional agriculture when resources are constrained. Figure 1.5 shows the threshold at which these technologies become cost-effective. The figure shows stylized cost curves for frontier farming and conventional farming. In the figure, R is a resource index that includes natural resources—such as space, water, nutrients, and arable land—and economic inputs—such as hard currency and feed or infrastructure inputs. It shows a direct advantage for frontier agricultural technologies when resources are constrained, or when R is less than or equal to the tipping point level of resources, or R^* ($R \leq R^*$). This is why these technologies

BOX 1.2 Farm-Level and Country-Level Surveys

This report carried out two surveys on insect farming in Africa between September 2019 and January 2020. The objective of the surveys, one at the country level and the other at the farm level, was a rapid assessment of insect farming activities in 13 African countries: Benin, Burundi, Cameroon, the Democratic Republic of Congo, the Arab Republic of Egypt, Ghana, Kenya, Madagascar, Rwanda, Tanzania, Tunisia, Uganda, and Zimbabwe. The World Bank has designated four of these countries—Burundi, Cameroon, the Democratic Republic of Congo, and Zimbabwe—as FCV (fragility, conflict, and violence) countries according to its list of fragile and conflict-affected situations,[a] although other surveyed subnational areas also suffer from FCV, such as Kenya's refugee camps. This was the first time that surveys on insect farming, not wild insect harvesting, were carried out in most of these countries. The *country-level survey* covered the history of insect farming, types of farms found, stakeholders involved, and challenges the industry faced in each country. The *farm-level survey* gathered information from 161 farms on farmer demographics, species farmed, production systems, farm outputs, insect markets, and benefits and challenges. A network of surveyors administered the surveys to insect farmers and farming experts in each country. In some cases, the surveyors recruited enumerators to administer the survey at the farm level. Using a nonrandomized method, the surveyors identified the survey's sample frame and, as such, the survey results are not a representative sample of insect farms in the surveyed countries. All the farmer quotes in the report were taken from the two surveys. As a final note, the two surveys are not statistically representative.

a. The *terms fragile and conflict-affected situations* and *FCV countries* are used interchangeably throughout the report.

METHODOLOGY

Each chapter relies on original research and a literature review on themes related to food security, insect farming, and hydroponics. The original research was particularly important in filling in large data and information gaps on these themes. The research team carried out two surveys—a country-level survey and a farm-level survey—on insect farming in 13 African countries (for details, see box 1.2). Map 1.1 shows the countries in which data were collected for this report. Chapter 4 models the frontier agricultural supply chain for black soldier fly larvae production in Africa and, individually, in Zimbabwe and 10 other African countries, to approximate protein production levels, waste reduction levels, and economic, climate, employment, and environmental benefits given different crop waste substrates. This research is supplemented with observations from field visits to Kenya, Zimbabwe, Thailand, and the Republic of Korea. The emergence of COVID-19 limited the team's ability to complete planned visits to other African FCV countries.

MAP 1.1 Countries in Which Insect Farming Data Were Collected for This Report

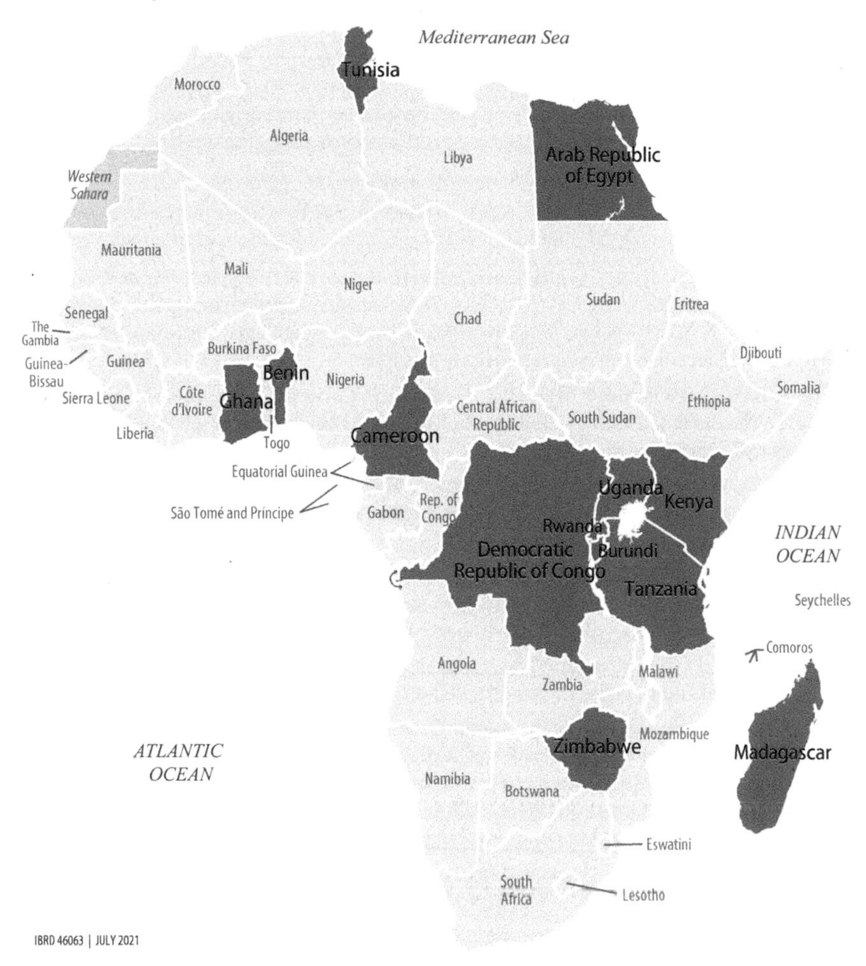

Source: Original map for this publication.

NOTES

1. The World Bank's fiscal year 2021 list of countries in fragile and conflict-affected situations in the Africa region includes Burkina Faso, Burundi, Cameroon, the Central African Republic, Chad, the Comoros, the Democratic Republic of Congo, the Republic of Congo, Eritrea, The Gambia, Guinea-Bissau, Liberia, Mali, Mozambique, Niger, Nigeria, Somalia, South Sudan, Sudan, and Zimbabwe. The list is available at http://pubdocs.worldbank.org/en/888211594267968803/FCSList-FY21.pdf.
2. World Food Programme reporting from January 24, 2021.
3. As of 2021, 20 African countries were classified as FCV, but undernourishment data were available for only 13 countries.

WMO (World Meteorological Organization). 2019. *State of the Climate in Africa 2019*. Geneva: WMO. https://library.wmo.int/doc_num.php?explnum_id=10421.

World Bank. 2020a. *World Development Indicators*. Washington, DC: World Bank.

World Bank. 2020b. *Addressing Food Loss and Waste: A Global Problem with Local Solutions*. Washington, DC: World Bank. https://openknowledge.worldbank.org/bitstream/handle/10986/34521/Addressing-Food-Loss-and-Waste-A-Global-Problem-with-Local-Solutions.pdf?sequence=1&isAllowed=y.

World Hunger Education Service. 2018. "Africa Hunger and Poverty Facts." Hunger Notes, updated August 2018. https://www.worldhunger.org/africa-hunger-poverty-facts-2018/.

CHAPTER TWO

Food Security Context

HIGHLIGHTS

- The number of undernourished people is increasing in Africa, particularly in countries affected by fragility, conflict, and violence (FCV). In the 10 African FCV countries for which data were available, there were 38.2 million undernourished people in 2001. This increased slightly to 39 million by 2012 and jumped dramatically to 55.5 million by 2018.
- In African FCV countries, poor nutrition contributes to poor health, with anemia rates as high as 57 percent among women in The Gambia and child stunting rates as high as 54 percent in Burundi. This is leading to cognitive disorders, undermining early childhood development, and eventually affecting economic development.
- African FCV countries depend on food imports to ensure food supplies. The 13 African FCV countries with available data imported 13.6 million tons of cereals in 2018 to meet the domestic food consumption demand.
- African FCV countries are overly dependent on agriculture for generating employment, which can undermine food security. In 2019, 13 African FCV countries employed over 50 percent of females in the agriculture sector. In Burundi, 93 percent of females work in agriculture.
- Conflict and violence continue to displace people at an increasing rate in Africa. The total number of refugees, asylum seekers, stateless individuals, and internally displaced persons increased from 18.5 million people in 2015 to 33.4 million in 2019 in Africa—an increase of 80.5 percent.

- Climate variability and change will continue to aggravate food insecurity in African FCV countries. Renewable freshwater resources declined in all African FCV countries from 2002 to 2017, and the number of undernourished people in drought-sensitive countries has increased by 45.6 percent since 2012.

The purpose of this chapter is to shed light on the food security situation in African FCV countries. It describes the contextual background in which any disruptive change from a linear economy to a circular economy would occur. The chapter is organized into five sections. The first section looks at food security outcomes in Africa and finds that the situation was dire even before the COVID-19 (coronavirus) pandemic, with high food insecurity and undernutrition. The second section assesses the food supply structure in these countries and finds that they rely heavily on food imports, particularly of proteins, and have high food production variability. The third section looks at the economic structure of the region's agriculture sector, finding that it has high potential for development and represents a disproportionately high share of the economy. The fourth section describes some of the demographic pressures in African FCV countries and finds that population growth and migration contribute to food security pressures. The fifth section shows how climate change is dwindling already scarce natural resources and will accelerate food security challenges.

There is a significant data gap for FCV countries in Africa. This is particularly true for time series data for some national economies. These data gaps limit the ability of analytical work, like this report, to provide a comprehensive diagnostic of socioeconomic conditions related to food security in these economies. This chapter mitigates these gaps by combining information from multiple sources, including the United Nations, the World Bank's World Development Indicators, the United Nations High Commissioner for Refugees, and the Food and Agriculture Organization of the United Nations (FAO), among other sources. The chapter presents and discusses data for 20 FCV countries in Africa: Burkina Faso, Burundi, Cameroon, the Central African Republic, Chad, the Comoros, the Democratic Republic of Congo, the Republic of Congo, Eritrea, The Gambia, Guinea-Bissau, Liberia, Mali, Mozambique, Niger, Nigeria, Somalia, South Sudan, Sudan, and Zimbabwe. These FCV countries are included in the Harmonized List of Fragile Situations released by the World Bank in 2021[1] and differ from the World Bank's FCV lists from previous years. History shows that conflict disrupts societies and their economies, but it also disrupts data collection and reporting; thus, it is not a surprise that the countries affected by FCV are the ones for which the most data are missing. This chapter provides the most current and available data and compares them with benchmarks from the early 2000s. In several cases, however, the most recent data are several years old; therefore, the chapter uses data averages to fill in missing information. Not all the tables and figures in the chapter contain all 20 African FCV countries because certain countries lack data on particular indicators. Indications in the text and footnotes provide details on data adjustments.

Dimensions of Food Security

Food security is a broad concept that includes four dimensions: availability, access, utilization, and stability (figure 2.1). "Food availability" is determined by production capabilities, inventory, and net trade. "Food access" refers to an individual's capability to obtain food that might be available but not accessible because of economic reasons—such as low incomes or high prices—or physical barriers—such as lack of transportation to reach food markets. "Food utilization" refers to individuals' ability to ensure that the food they access meets their nutritional and physiological needs. Factors that determine this include nutrition, health outcomes, food quality, food diversity, and feeding practices. "Food stability" is an individual's ability to obtain food over time. Food availability, access, utilization, and stability may not always be guaranteed, especially in FCV countries. All four dimensions must be met simultaneously to ensure food security (Martin-Shields and Stojetz 2018; FAO 2008).

FIGURE 2.1 The Four Dimensions of Food Security

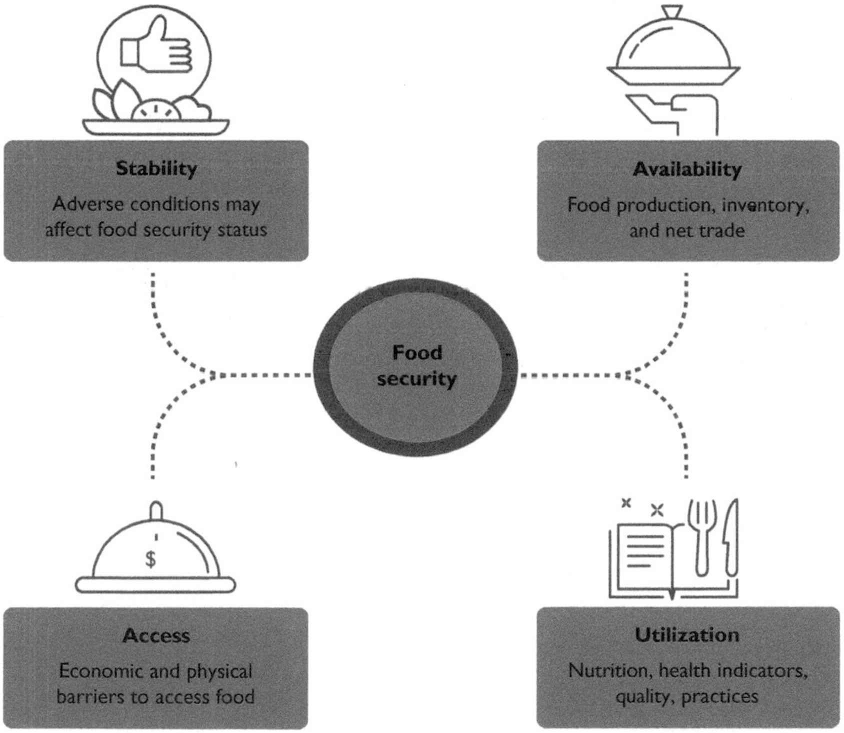

Source: Adaptation from FAO 2008.

The four dimensions of food security are measured by a myriad of alternative but complementary indicators. For instance, food utilization can be measured using the FAO's indicator on the prevalence of undernourishment, which refers to the population whose dietary food consumption is below a threshold that would ensure dietary energy levels for a normal active and healthy life (FAO and ECA 2018). Undernourishment can be measured by health outcomes such as stunting, anemia, underweight population, and other health indicators. Food supply can be measured by prices and food availability. These indicators and others help provide this chapter's comprehensive overview of food security in African FCV countries.

Food insecurity is a negative feedback loop. Undernourishment causes anemia and stunting, leading to impairment, poor mental development, and lower cognitive function, and hinders a person's physical capacity to perform work and earn wages. These effects create a negative feedback loop since lower wages and unemployment contribute to poverty and food insecurity, further reducing a person's cognitive development and human capital accumulation (FAO and ECA 2018; FAO 2008; World Bank 2006b). This, in turn, reduces productivity and increases poverty (see figure 2.2).

Conflict and violence add another complex layer to food insecurity's negative feedback loop. An FAO study finds "that causal and substantive linkages exist between food security and violent conflict, spanning the individual, local, regional, country and global levels" (Martin-Shields and Stojetz 2018, 23). Conflict negatively affects economic activity and disrupts transportation and access to markets, which particularly affects smallholder farmers who are cut off from inputs and consumer markets. Koren and Bagozzi (2017, 351) provide theoretical and empirical evidence linking violent conflict to food insecurity, finding that "the existence of immediate conflict in a region leads armed actors to discount the benefits of future interactions in favor of obtaining food immediately, using violence if necessary." Moreover, there is evidence that countries with more cropland are more susceptible to conflict because of the strategic needs of warring factions to obtain food (Koren and Bagozzi 2017).

Undernourishment

Food insecurity is more prevalent in countries marked by FCV. On the African continent, approximately 53 percent of the population experiences food insecurity. However, food insecurity tends to be much worse in FCV countries. During 2016–18, 86.2 percent of Liberia's population and 86.6 percent of Mozambique's population experienced severe or moderate food insecurity (FAO et al. 2019). The World Bank classifies each of these countries as FCV, suggesting that food insecurity is particularly acute in FCV countries, where conflicts disrupt economic activity, access to food, and production (Martin-Shields and Stojetz 2018).

FIGURE 2.2 Food Insecurity's Negative Feedback Loop

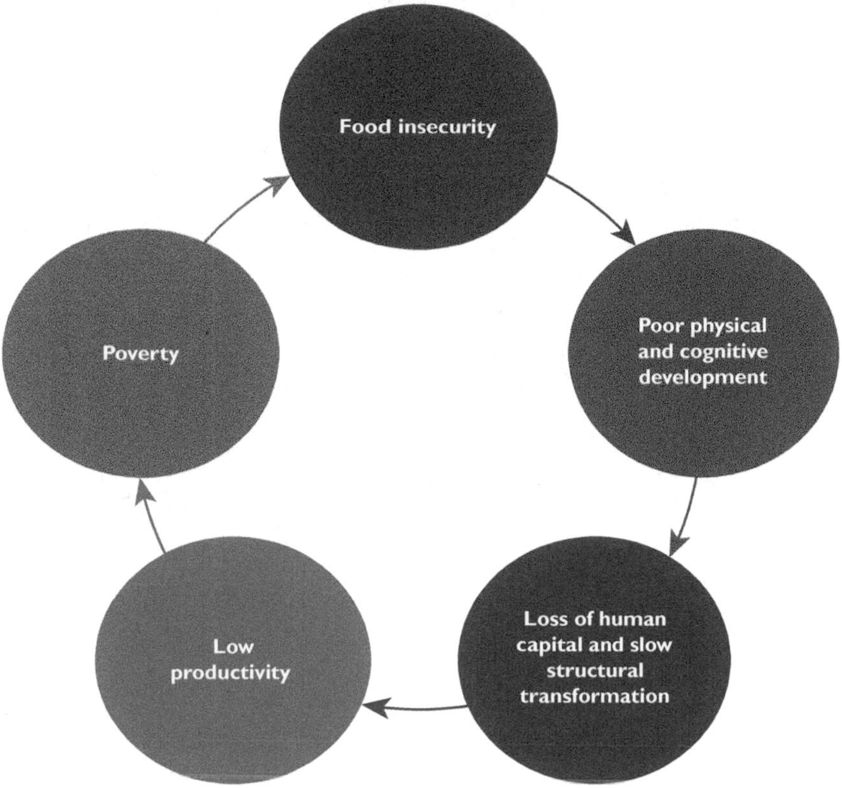

Source: Adaptation from FAO 2008 and World Bank 2006a.

The number of undernourished people is increasing on the African continent, particularly in FCV-affected countries. In 2018, there were 55.5 million undernourished people in the 10 African FCV countries for which data were available, in 2012 there were 39 million, and in 2001 there were 38.2 million. This represents an increase of 16.5 million people during the six-year period from 2012 to 2018. From 2001 to 2012, the number of undernourished people in these 10 FCV[2] countries increased by only 0.8 million people, despite the total population increasing from 287.5 million to 370.2 million—a 36 percent increase—during that time. However, from 2012 to 2018, the undernourished population increased from 370.2 million to 437.9 million—a 42 percent increase—while the population increased by only 18 percent.

In Africa, FCV countries are less food secure and subject to higher undernourishment rates than non-FCV countries (figure 2.3). In 2018, 20.5 percent of the population in African FCV countries was undernourished, compared

FIGURE 2.3 Undernourishment Rates in FCV versus Non-FCV Countries in Africa, 2001–18

Percent

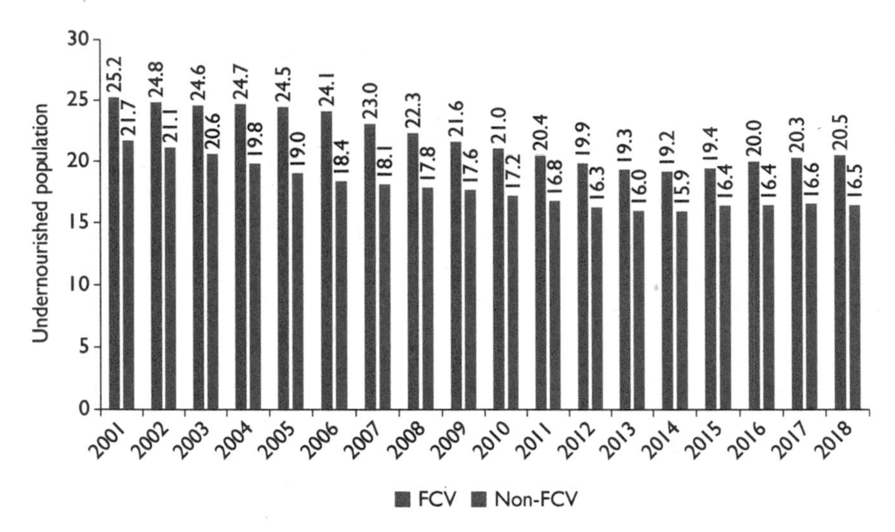

Source: Compilation for this publication, based on data from World Bank 2020.
Note: The fragility, conflict, and violence (FCV) calculations are for the 10 FCV countries for which data are available.

with 16.5 percent of the population in African non-FCV countries. The FAO estimates that globally, 60 percent of the 815 million undernourished individuals and 79 percent of the 155 million stunted children live in countries affected by violent conflict (FAO et al. 2017).

Undernourishment from food insecurity was high in each of the 10 African FCV countries even before the COVID-19 pandemic. In 2018, Nigeria had the most undernourished people, with 24.7 million, followed by Mozambique, 9.6 million; Chad, 6.1 million; Sudan, 5.2 million; Burkina Faso, 3.8 million; Liberia, 1.8 million; Cameroon, 1.6 million; the Republic of Congo, 1.5 million; Mali, 1.0 million; and The Gambia, 0.3 million (figure 2.4). Conflict in the Central African Republic led to increased undernourishment when it disrupted production and caused food prices to skyrocket. The FAO (2018) estimates that 34 percent of the Central African Republic's population needed urgent food assistance in early 2018. Evidence also suggests that in 2018, conflict disrupted food access and production in Burundi, Chad, the Democratic Republic of Congo, and Mali. For these statistics, Burundi, the Central African Republic, and the Democratic Republic of Congo, among other African FCV countries, often lack data.

Undernourishment has not improved significantly in most African FCV countries over the past two decades. Between 2001 and 2018, the number of

FIGURE 2.4 Undernourishment Is Pervasive and Increasing among FCV Countries

Millions of people

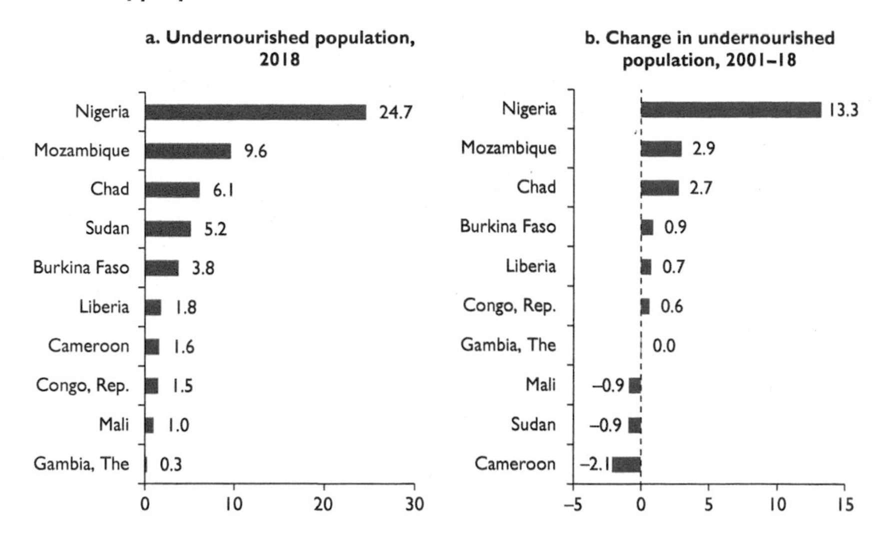

Source: Compilation for this publication, based on data from the Food and Agriculture Organization of the United Nations.
Note: FCV = fragility, conflict, and violence.

undernourished people decreased in Mali, Sudan, and Cameroon. However, other countries experienced an increase in undernourishment, sometimes considerably. From 2001 to 2018, the number of undernourished people increased by 13.3 million in Nigeria, 2.9 million in Mozambique, 2.7 million in Chad, 0.9 million in Burkina Faso, 0.7 million in Liberia, and 0.6 million in the Republic of Congo (figure 2.4). In The Gambia, the number of undernourished people stayed the same but the proportion of undernourished people decreased from 18 percent in 2001 to 11.9 percent in 2018. The Gambia aside, the general trend of increased undernourishment indicates that food systems in African FCV countries are neither working nor capable of meeting the food needs of local populations.

Health Outcomes

Female anemia rates, caused by malnutrition,[3] are severe in African FCV countries. Anemia, or iron deficiency, rates are considered severe in a country if over 40 percent of the population is anemic. A lack of nutrients in the diet is one cause of anemia. This mainly results from a lack of iron in the diet, but a lack of folate, vitamin B12, or vitamin C can also contribute (Brazier 2019). Anemia levels are severe among women in 13 of the 20 African FCV countries (figure 2.5). The literature shows that anemia is associated with stunting,

FIGURE 2.5 Prevalence of Anemia in Women of Reproductive Age (15–49 Years), 2016

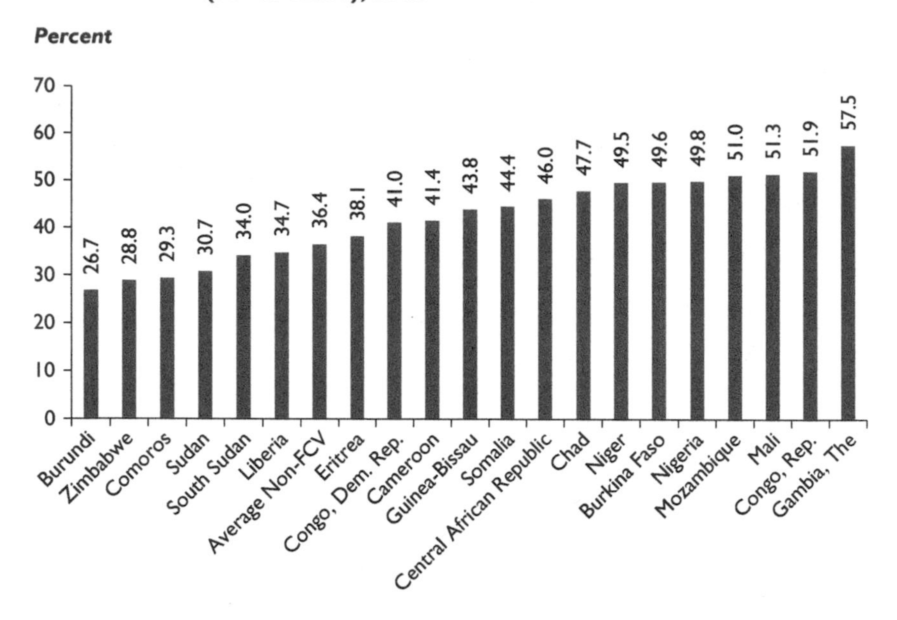

Source: Compilation for this publication, based on data from World Bank 2020.
Note: FCV = fragility, conflict, and violence.

low birth weight, early delivery, high mortality rates among women, and deficient brain development in babies, which leads to poor neurocognitive outcomes (World Bank 2017; Viteri 1994).

Among African FCV countries, anemia rates range from 26.7 percent in Burundi to 57.5 percent in The Gambia. More than half of the women of childbearing age suffer from anemia in Mozambique (51.0 percent), Mali (51.3 percent), the Republic of Congo (51.9 percent), and The Gambia (57.5 percent). Anemia rates are slightly lower but still severe among women in Nigeria, Burkina Faso, Niger, Chad, the Central African Republic, Somalia, Guinea-Bissau, Cameroon, and the Democratic Republic of Congo, ranging from 41.0 to 49.8 percent. Investments to reduce anemia are associated with increased productivity and higher economic returns. A World Bank (2017) report estimates that each dollar invested to reduce anemia generates US$12 in economic returns.

Poor nutrition has resulted in high stunting rates in African FCV countries, leading to health and economic losses. According to the World Health Organization, inadequate nutrition, caused by not eating enough or by eating foods that lack growth-promoting nutrients, is the most direct cause of stunting in children (WHA, n.d.). Stunting impairs children's cognitive and physical development, limiting their ability to learn and reducing their future income and productivity as adults (FAO and ECA 2018). The economic impacts of stunting are significant for individuals and the economy as a whole. The per capita income penalty from stunting in Africa is 9 to 10 percent (Galasso and Wagstaff 2018), and children who escape stunting are 33 percent more likely to escape poverty as adults (Hoddinott et al. 2008; Horton and Steckel 2013). Moreover, reductions in stunting can increase gross domestic product (GDP) by 4 to 11 percent in Asia and Africa (Shekar et al. 2017). Map 2.1 shows that stunting rates in Africa are among the most severe in the world, with approximately one-third of children under five having a height for their age that is more than two standard deviations below the international median.

In African FCV countries, 32 million children under five years of age are stunted (see table 2A.2). Figure 2.6 shows that stunting rates were higher than 40 percent in Burundi (54.2 percent), Eritrea (52.5 percent), Niger (48.5 percent), the Democratic Republic of Congo (42.7 percent), Mozambique (42.3 percent), and the Central African Republic (40.8 percent). Stunting rates were between 30 and 40 percent in Chad (39.8 percent), Sudan (38.2 percent), Nigeria (36.8 percent), South Sudan (31.3 percent), the Comoros (31.1 percent), and Liberia (30.1 percent). Cameroon, Guinea-Bissau, Mali, Somalia, Burkina Faso, Zimbabwe, the Republic of Congo, and The Gambia all had stunting rates under 30 percent. High stunting rates are associated with long-term negative impacts on cognitive and physical development, which can adversely affect human capital accumulation and economic growth (Skoufias 2018).

FOOD SUPPLY

African FCV countries rely on food imports to ensure food supplies. Conflict, forced displacement, and low agricultural productivity keep African FCV countries from being self-sufficient in food production. The 13 African FCV countries with available data imported 13.6 million tons of cereals, including 7.8 million tons of wheat and 690,000 tons of maize in 2018, to meet domestic food consumption demand (table 2.1). Imports of cereals were particularly large in Nigeria (6.392 million tons), Mozambique (1.619 million tons), Cameroon (1.582 million tons), Burkina Faso (0.652 million tons), and the Republic of Congo (0.482 million tons). In contrast, cereal exports were negligible across these countries. Although there are benefits to relying on food imports to ensure food security, there are also major risks. African FCV countries often have poor infrastructure, weak institutions, and limited resources to pay for food imports, and therefore cannot ensure that import flows are not disrupted. Furthermore, countries that rely heavily on food imports may still fall short of demand. For example, Nigeria experienced a cereal deficit of 341,000 tons in 2018 despite importing nearly 6.4 million tons of cereals. Other African FCV countries with cereal deficits in 2018 included Zimbabwe (−217,000 tons), Mozambique (−133,000 tons), and Liberia (−16,000 tons).

Animal source foods (ASF), like meat and dairy, are key sources of protein for people in African FCV countries.[4] Seafood and fresh vegetables represent a

MAP 2.1 Stunting Rates in Low- and Middle-Income Countries, 2015

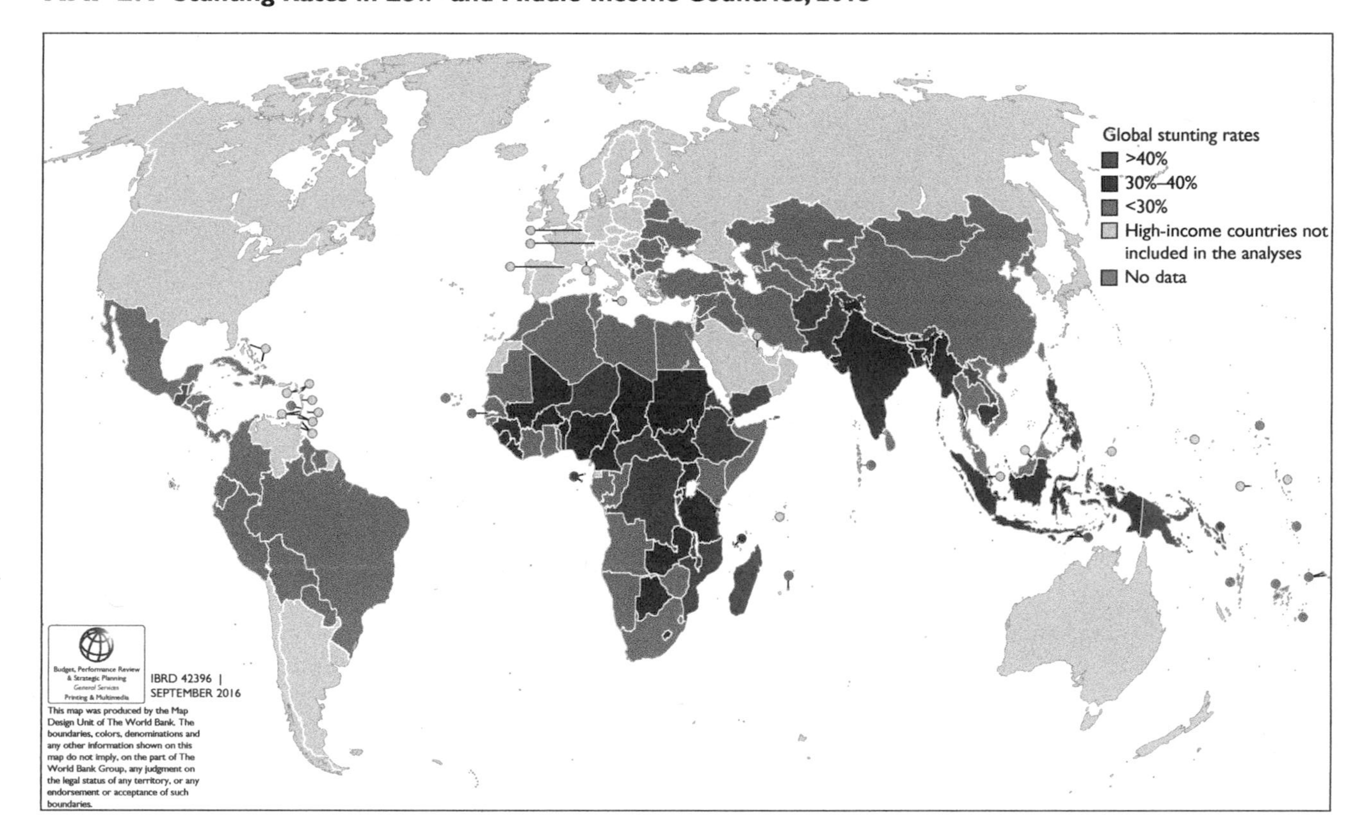

Source: World Bank 2017.

 **Prevalence of Stunting, or a Height-for-Age More Than Two Standard Deviations below the International Median, among Children Younger Than Five Years**

Percent

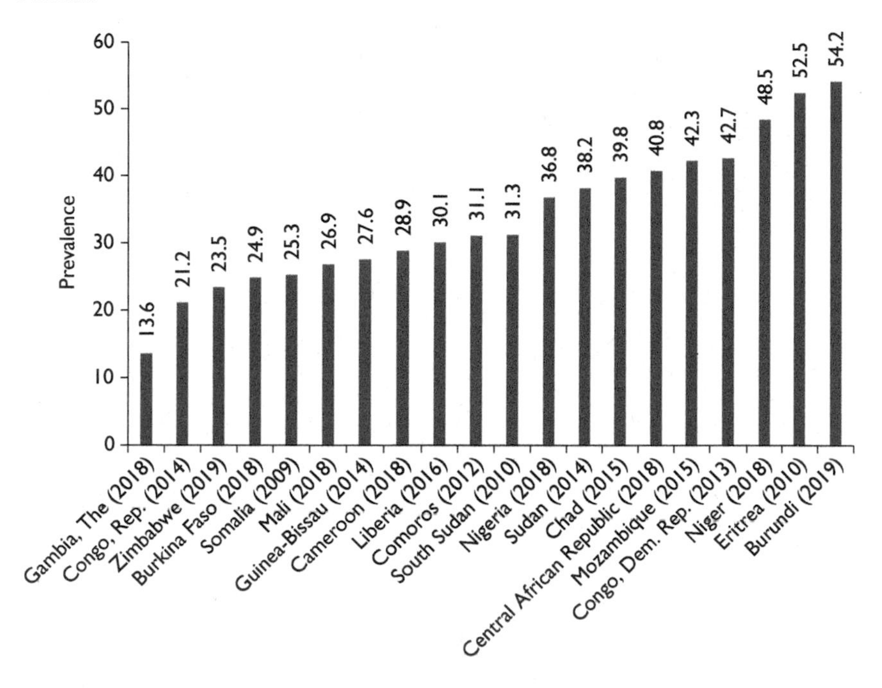

Source: Compilation for this publication, using data from World Bank 2020.
Note: The values are from the most recently available data (years are shown in parentheses).

smaller portion of the food supply (figure 2.7). Mali stands out with an average supply of animal products at 336 calories (kcals) per capita per day, followed by the Central African Republic at 252 kcal/capita/day and the Republic of Congo at 216 kcal/capita/day, all of which are higher than the African average of 215 kcal/capita/day. By contrast, Nigeria's supply of animal products is only 84 kcal/capita/day. Vegetable supplies are even lower in these countries. The Central African Republic and the Republic of Congo produce only 11 and 41 kcal/capita/day of vegetables, respectively. Burkina Faso produces only 12 kcal/capita/day and Chad produces the fewest vegetables at 5 kcal/capita/day. Meanwhile, Guinea-Bissau and Niger have the lowest fish and seafood supply, with only 3 and 4 kcal/capita/day, respectively. These figures suggest that African FCV countries lack the food production diversity and food subsectors to ensure balanced, nutritious diets.

TABLE 2.1 Cereal Balance Sheet for 13 African FCV Countries with Available Data, 2018

Thousands of tons

	Burkina Faso	Cameroon	Central African Republic	Chad	Congo, Rep.	Gambia, The	Guinea-Bissau	Liberia	Mali	Mozambique	Niger	Nigeria	Zimbabwe	FCV total
Production	4,991	3,950	141	3,022	30	222	214	258	10,160	1,513	6,100	26,872	1,423	58,896
Imports	652	1,582	5	116	482	300	177	402	470	1,619	928	6,392	440	13,565
Exports	6	15	0	0	1	0	0	8	17	21	436	27	9	540
Total supply	**5,637**	**5,517**	**146**	**3,138**	**511**	**522**	**391**	**652**	**10,613**	**3,111**	**6,592**	**33,237**	**1,854**	**71,921**
Consumption	4,288	3,615	121	2,551	450	395	309	612	4,472	2,934	4,939	26,706	1,772	53,164
Feed	121	919	0	52	19	19	10	20	1,091	120	799	4,011	10	7,191
Seed	117	82	8	43	0	10	9	10	208	72	266	369	49	1,243
Processing	669	177	6	7	48	9	6	2	57	83	3	494	84	1,645
Other (nonfood)	0	163			0	20			1,782	1	0	57		2,023
Total demand	**5,195**	**4,956**	**135**	**2,653**	**517**	**453**	**334**	**644**	**7,610**	**3,210**	**6,007**	**31,637**	**1,915**	**65,266**
Losses	226	471	6	340	1	20	9	32	763	88	816	1,945	155	4,872
Residual	0	0	0	0	−17	0	0	−7	1,563	−56	−434	−3	0	1,046
Stock variation	216	91	5	145	10	48	49	−16	676	−133	202	−341	−217	735

Source: Compilation for this publication, using data from the Food and Agriculture Organization of the United Nations.

Note: The numbers do not always add up because of rounding. FCV = fragility, conflict, and violence.

 **Food Supply in 13 African FCV Countries, 2018**

Kcal per capita per day

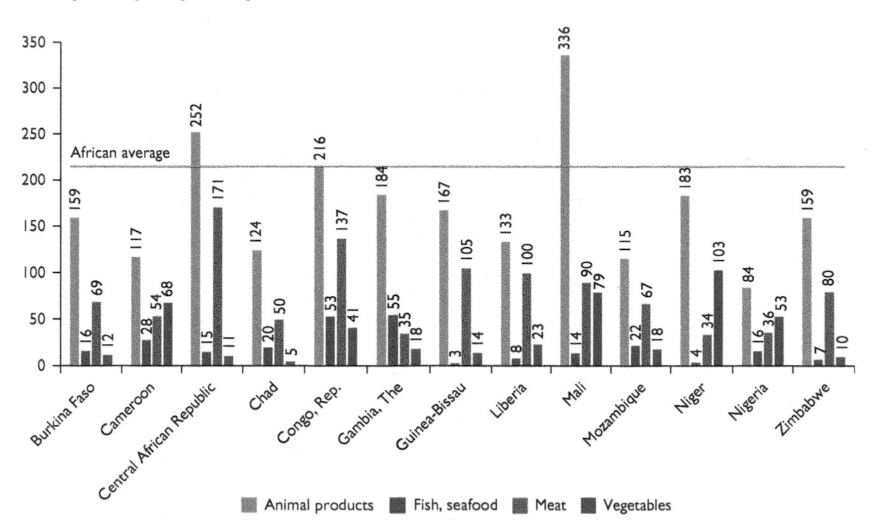

Source: Compilation for this publication, using data from the Food and Agriculture Organization of the United Nations.
Note: FCV = fragility, conflict, and violence; kcal = calories.

There are significant variations in the pattern and composition of food supplies in African FCV countries. Over the five-year period from 2014 to 2018, 7 of 13 African FCV countries experienced an increase in ASF supplies, or animal products, which include the subcategories of meat and fish, seafood (figure 2.8). Mali and the Central African Republic experienced the largest increases in ASF supply at 21 kcal/capita/day. Vegetable supplies increased by 30 kcal/capita/day in Mali and 12 kcal/capita/day in Niger, the most among African FCV countries, while the rest of the countries had small or no changes in vegetable supplies. Cameroon, Nigeria, and Zimbabwe all experienced declines. This is troublesome because animal products alone cannot ensure an adequate, balanced, and nutritious diet. These variations can be explained, in part, by disruptions in agricultural production and food supply systems caused by climate shocks, violence and conflict, low government capacity, inadequate or outdated technology, and the low skill levels of agricultural producers (IFPRI 2015).

Several African FCV countries fall below generally recommended protein intake standards. Dietary guidelines suggest that a 64-kilogram individual needs to consume, on average, 51 grams of protein per day. But the average daily per capita protein consumption is only 45 grams in the

FIGURE 2.8 Changes in Food Supply in 13 African FCV Countries, 2014–18

Kcal per capita per day

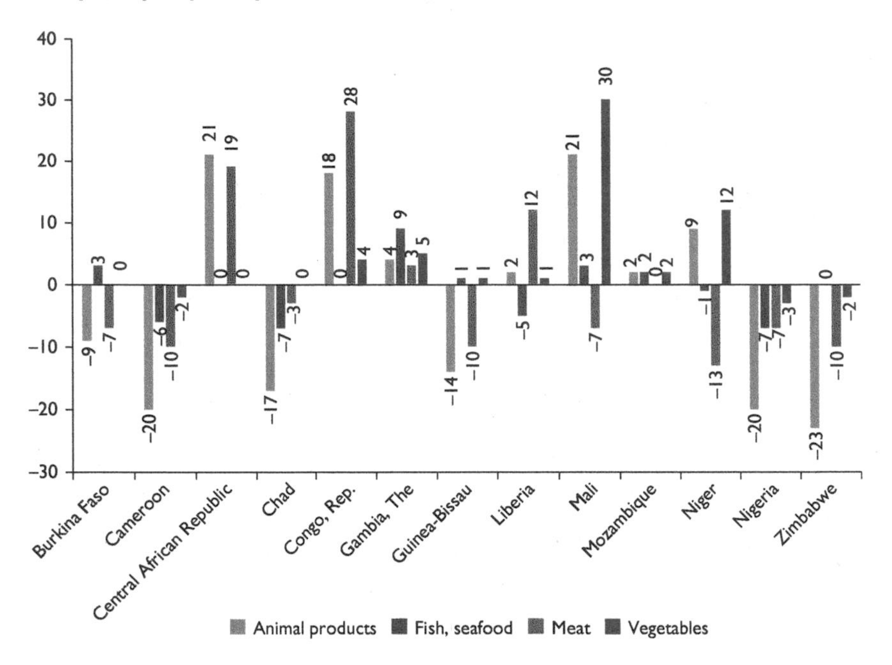

Source: Compilation for this publication, using data from the Food and Agriculture Organization of the United Nations.

Note: FCV = fragility, conflict, and violence; kcal = calories.

Central African Republic, 44 grams in Zimbabwe, 43 grams in Guinea-Bissau and Liberia, and 42 grams in Mozambique (figure 2.9). A United Nations Development Programme report finds that improvements in cereal and meat productivity are the most important determinants of protein supply in Africa (UNDP 2012). The report estimates that meat has the highest return on investment for increasing protein supplies in Guinea-Bissau, The Gambia, and the Democratic Republic of Congo. Cereal crops have the highest return in Burundi, and root crops have the highest return in Mali. Mali has the highest average daily protein supply at 84 grams per capita. This is because the diet in Mali is composed mostly of animal products, which are rich in protein.

Figure 2.10 shows that food production variability has increased in most African FCV countries over the past two decades. This variability means more volatility and instability in national food systems, undermining food security and limiting the revenue potential of food exports. Food production variability

FIGURE 2.9 Average Protein Supply in 13 African FCV Countries, 2018

Grams per capita per day

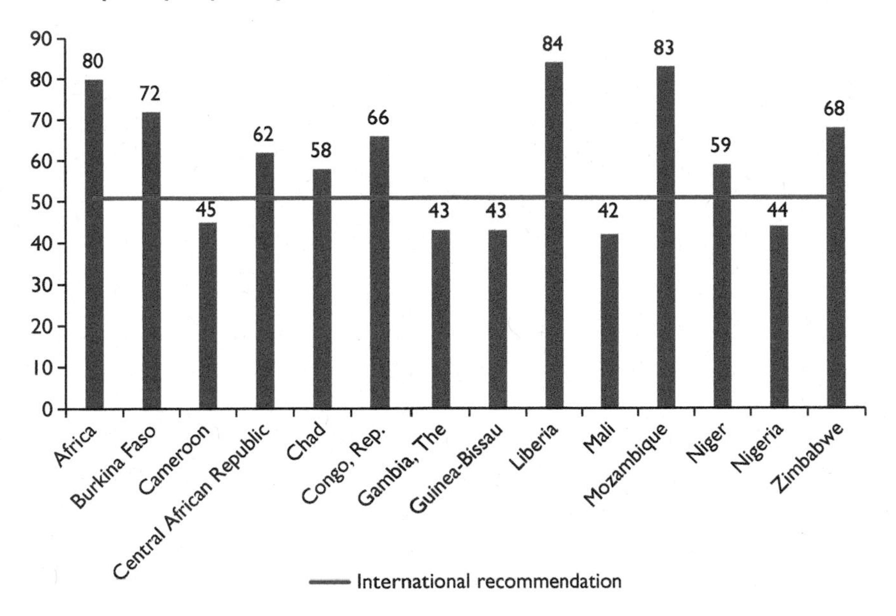

Source: Compilation for this publication, using data from the Food and Agriculture Organization of the United Nations.
Note: FCV = fragility, conflict, and violence.

also causes food price volatility, which leads to significant hardship for consumers, particularly the poor and smallholder farmers who cannot protect themselves from these price fluctuations.

Food exports from African FCV countries are important for generating foreign currency revenues. Food product exports account for a significant share of total merchandise exports in most African FCV countries. For example, food exports represented over 50 percent of total merchandise exports in The Gambia, the Comoros, Niger, and Burundi during 2017–19 (figure 2.11, panel a), which contrasts with the general decline in food as a proportion of total merchandise exports from 2000–02 to 2017–19 (figure 2.11, panel b). Over that period, the food contributions to total merchandise exports increased significantly in Niger, Mali, and Cameroon. Food exports represent a relatively small percentage of merchandise exports in Nigeria (1.9 percent), the Central African Republic (0.8 percent), and the Republic of Congo (0.3 percent). This is not surprising because the Central African Republic and the Republic of Congo, in particular, can barely produce enough protein to meet domestic

FIGURE 2.10 Change in Per Capita Food Production Variability in 18 African FCV Countries, 2000–16

Constant 2004–06 thousands of international dollars

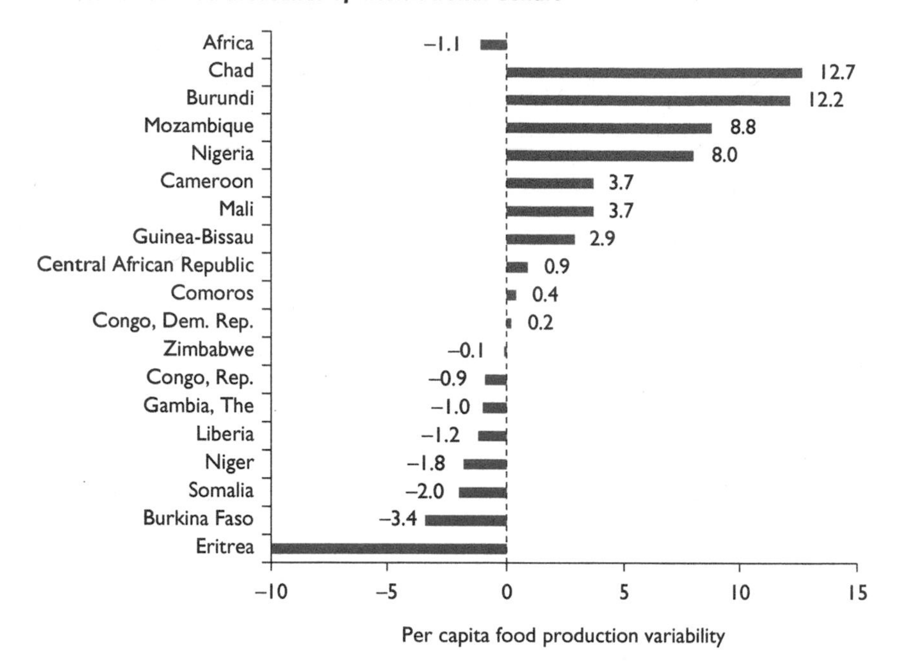

Source: Compilation for this publication, using data from World Bank 2020.
Note: FCV = fragility, conflict, and violence.

demand; thus, food exports are not a viable option for these countries to generate foreign currency revenue.

ECONOMIC STRUCTURE OF THE AGRICULTURE SECTOR

African FCV countries are overly dependent on agriculture for generating employment, which can undermine food security. In 2019, the agriculture sector employed 78.1 percent of males and 93.7 percent of females in Burundi, and 76.5 percent of males and 73.4 percent of females in Chad. Overall, 11 African FCV countries employed over 50 percent of males and 13 African FCV countries employed over 50 percent of females in the agriculture sector (table 2.2). These figures indicate a lack of economic diversification and opportunities beyond agriculture. They also imply that most

FIGURE 2.11 Food Exports as a Percentage of Merchandise Exports

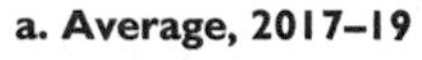

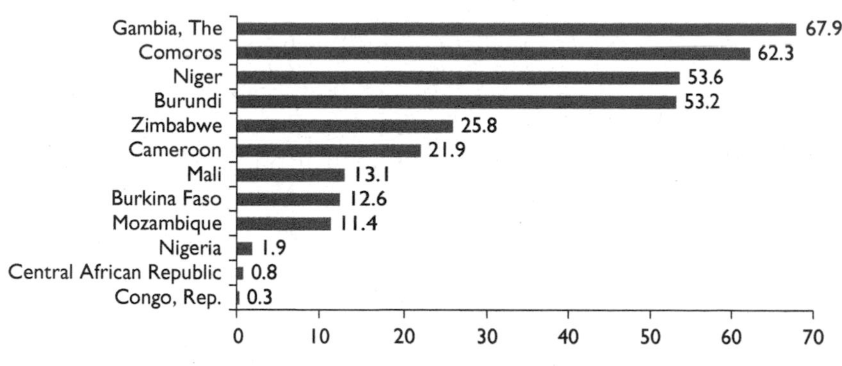

b. Change, 2000–02 to 2017–19

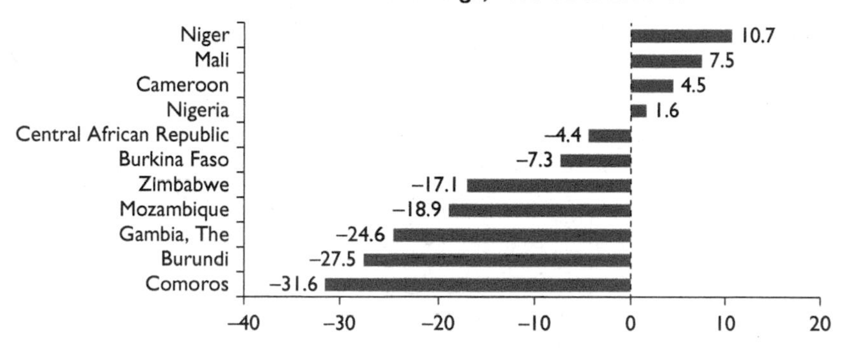

Source: Compilation for this publication, using data from World Bank 2020.
Note: The most recent data points for the Comoros and Sudan are for 2014.

people in African FCV countries are highly exposed to climate shocks or other risks inherent in agriculture, such as price volatility. A report by the FAO indicates that diversification and growth of nonfarm activities are critical for agricultural growth and to increase farmers' incomes (FAO 2014). Studies also suggest that food security requires integrating food and agricultural systems with rural and urban markets (Timmer 2017).[5] However, this structural transformation remains elusive for African FCV countries. That said, a strong and productive agriculture sector can improve livelihoods and reduce poverty "because profitable and productive farming is a catalyst in many rural communities for driving people on to better jobs, higher wages and an improved quality of life" (World Bank 2018a, 6). And an underdeveloped agriculture sector helps "fragility to emerge and persist" (ACP-EU 2018, 18).

TABLE 2.2 Agricultural Employment as a Percentage of Total Employment in African FCV Countries, 2000–19

Country	2000		2019	
	Male	Female	Male	Female
Burkina Faso	83.9	76.3	30.1	21.4
Burundi	86.4	96.6	78.1	93.7
Cameroon	63.1	70.0	39.8	47.7
Central African Republic	71.7	79.9	67.5	72.6
Chad	80.9	83.9	76.5	73.4
Comoros	52.8	66.1	36.3	31.6
Congo, Dem. Rep.	65.5	81.3	57.4	71.5
Congo, Rep.	40.4	44.3	35.0	32.0
Eritrea	61.4	75.9	58.7	68.3
Gambia, The	28.0	46.6	22.6	33.1
Guinea-Bissau	61.2	74.9	57.1	64.1
Liberia	50.0	54.1	44.7	40.3
Mali	71.3	74.3	62.3	62.6
Mozambique	70.8	91.3	59.8	79.8
Niger	77.4	75.7	74.8	69.5
Nigeria	53.8	42.9	44.5	23.6
Somalia	81.6	88.8	79.2	83.9
South Sudan	52.6	80.0	48.2	73.2
Sudan	46.8	65.1	33.6	51.8
Zimbabwe	52.1	69.9	62.8	69.5

Source: Compilation for this publication, using data from World Bank 2020.
Note: FCV = fragility, conflict, and violence.

On average, fishing, forestry, and agriculture contribute more to the economies of African FCV countries than to those of African non-FCV countries. Liberia, which is still largely an agriculture-based economy, experienced the largest decline in the share of value added in agriculture, forestry, and fishing in GDP, dropping from 76.1 percent in 2000 to 39.1 percent in 2019, a 37.0-percentage-point decline in the contribution of agriculture to GDP (table 2.3). The contribution of agriculture to GDP also declined significantly in Sudan (from 40.7 percent in 2000 to 11.6 percent in 2019), Burundi (from 44.1 percent in 2000 to 28.9 percent in 2019), and the Democratic Republic of Congo (from 32.0 percent in 2000 to 20.0 percent in 2019).

TABLE 2.3 Value Added from Agriculture, Forestry, and Fishing as a Percentage of GDP in 19 African FCV Countries, 2000–19

Country	2000	2010	2019	Change (2000–19)
Burkina Faso	24.9	24.1	20.2	–4.7
Burundi	44.1	38.4	28.9	–15.2
Cameroon	16.7	14.1	14.5	–2.2
Central African Republic	—	37.4	32.4	—
Chad	40.7	51.9	42.6	1.9
Comoros	29.4	30.4	33.1	3.7
Congo, Dem. Rep.	32.0	21.4	20.0	–12.0
Congo, Rep.	5.3	4.0	7.8	2.5
Eritrea	12.6	—	—	—
Gambia, The	24.5	35.2	21.8	–2.7
Guinea-Bissau	41.7	45.1	52.5	10.8
Liberia	76.1	44.8	39.1	–37.0
Mali	32.9	33.0	37.3	4.4
Mozambique	19.1	26.8	26.0	6.9
Niger	36.6	35.8	37.8	1.2
Nigeria	21.4	23.9	21.9	0.5
South Sudan	—	5.3	—	—
Sudan	40.7	23.3	11.6	–29.1
Zimbabwe	15.7	9.6	8.3	–7.4
Average, African FCV countries	30.3	28.0	26.8	–4.9[a]
Average, African non-FCV countries	19.6	16.7	14.8	–4.8

Source: Compilation for this publication, using data from World Bank 2020.
Note: FCV = fragility, conflict, and violence; GDP = gross domestic product; — = not available.
a. Excludes countries without data from 2000, 2019, or both.

Agricultural labor productivity is low in African FCV countries, posing a barrier to food security and income generation. The average annual output per worker in the primary sector of African FCV countries is lower than the average for African non-FCV countries, which was Int$12,339 in 2018. More specifically, the average annual output per worker in the primary sector was less than Int$1,000 in Burundi (Int$602 in 2017), Zimbabwe (Int$826 in 2014), and

Mozambique (Int$960 in 2015). In contrast, the African FCV countries with the highest average annual output per worker in the primary sector were Nigeria (Int$11,128 in 2019), the Comoros (Int$9,813 in 2014), and Sudan (Int$9,668 in 2011) (figure 2.12). The overall low performance and labor productivity of the primary sector has negative impacts on food production, food security, and income creation. In addition, low agricultural productivity suggests that food systems in these countries have underdeveloped production, processing, and distribution practices. It is likely that conflict and violence have hindered farmers in these countries from adopting modern farming methods, which hampers the sector's growth and the ability of these economies to address pressing food and economic insecurity. Along with low productivity more generally, this has ramifications for income levels, which have largely not increased. Poverty rates in the 18 African FCV countries (listed in table 2.4) range from 37.5 percent in Cameroon to 76.4 percent in South Sudan; 16 of the 18 countries have poverty rates above 40 percent.

FIGURE 2.12 Average Annual Output per Worker in Agriculture, Forestry, and Fishery in 13 African FCV Countries, Various Years

International dollars, 2017 purchasing power parity

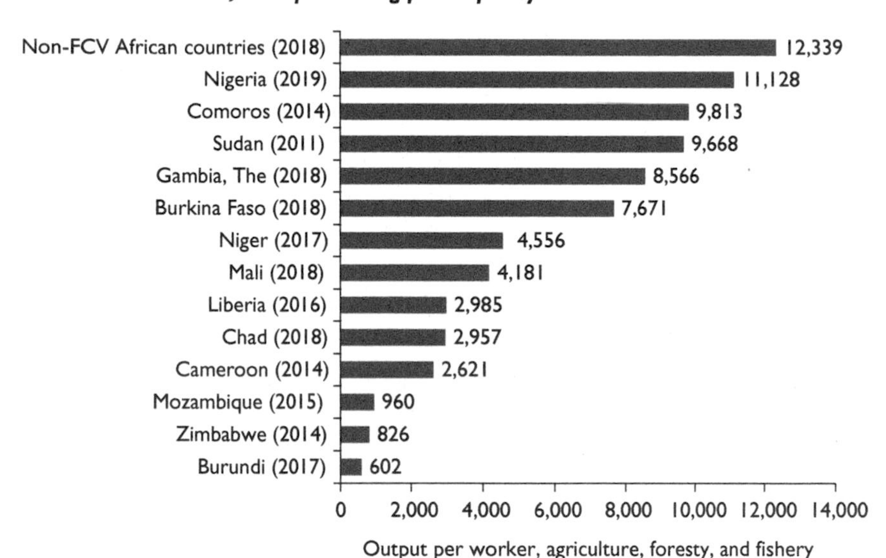

Source: Compilation for this publication, using data from World Bank 2020.
Note: The data for South Sudan and Burundi are from 2016. FCV = fragility, conflict, and violence.

TABLE 2.4 Population Living below the National Poverty Line in 18 African FCV Countries, Various Years

Country	Percent	Year
Burkina Faso	41.4	2019
Burundi	64.9	2013
Cameroon	37.5	2014
Central African Republic	62.0	2008
Chad	42.3	2018
Comoros	42.4	2013
Congo, Dem. Rep.	63.9	2012
Congo, Rep.	40.9	2011
Gambia, The	48.6	2015
Guinea-Bissau	69.3	2010
Liberia	50.9	2016
Mali	42.1	2019
Mozambique	46.1	2014
Niger	40.8	2018
Nigeria	40.1	2018
South Sudan	76.4	2016
Sudan	46.5	2009
Zimbabwe	38.3	2018

Source: Compilation for this publication, using data from World Bank 2020.
Note: FCV = fragility, conflict, and violence.

POPULATION CHANGE IN FCV COUNTRIES

Population growth varies in African FCV countries. While population levels consistently rise in the rest of Africa and the world, FCV countries see more variability, likely because of greater fragility, conflict, and forced migration. From 2000 to 2019, the rate of population growth declined in 11 African FCV countries (Chad, Mali, The Gambia, Somalia, Cameroon, the Republic of Congo, Liberia, Sudan, the Comoros, the Central African Republic, and South Sudan) and increased in eight countries (Niger, the Democratic Republic of Congo, Burundi, Burkina Faso, Mozambique, Nigeria, Guinea-Bissau, and Zimbabwe) (figure 2.13). Conflict-plagued and agriculture-based South Sudan experienced the most significant decline in population growth, from 3.9 percent per year in 2000–04 to less than 1 percent in 2015–19. Population

FIGURE 2.13 Average Annual Population Growth in 19 African FCV Countries from 2000–04 to 2015–19

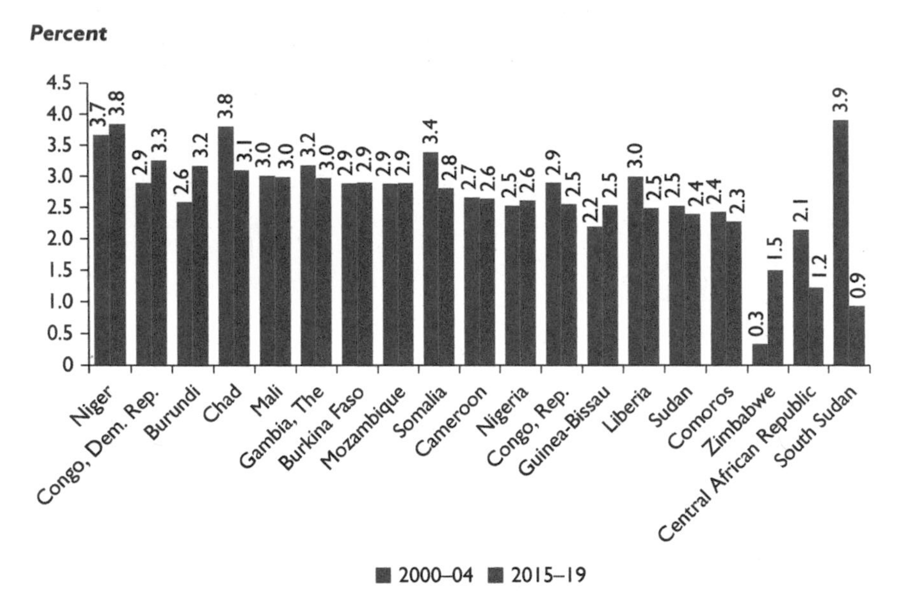

Source: Compilation for this publication, using data from World Bank 2020.
Note: FCV = fragility, conflict, and violence.

growth in the Central African Republic decreased from 2.1 percent in 2000–04 to 1.2 percent in 2015–19. Zimbabwe, by contrast, had the highest population growth, jumping from 0.3 percent in 2000–04 to 1.5 percent in 2015–19. High population growth in countries that are already food insecure can lead to even greater food insecurity (Population Action International 2011).

Most people in African FCV countries continue to live in rural areas despite the increase in rural-to-urban migration. While African non-FCV countries have slightly more than 50 percent of their populations living in urban areas on average, only five of 19 African FCV countries have majority urban populations, but this proportion is growing (figure 2.14). The relative growth of the urban population is often associated with a process of structural transformation in cities where services, job opportunities, and manufacturing industries are developing rapidly. However, this urbanization is unplanned and cities often lack the capacity to absorb incoming populations, resulting in overcrowding and congestion, and thereby exacerbating socioeconomic problems such as food insecurity. Nigeria, Mali, The Gambia, Burkina Faso, Somalia, and Cameroon had the most significant increases in urban population from 2000 to 2019. Urbanization has been slow in the Comoros, Chad, Sudan, South Sudan, the Central African Republic, and Burundi. Zimbabwe is the only African FCV country in which urbanization is not taking place—the

FIGURE 2.14 Population Living in Urban Areas in African FCV Countries, 2000 and 2019

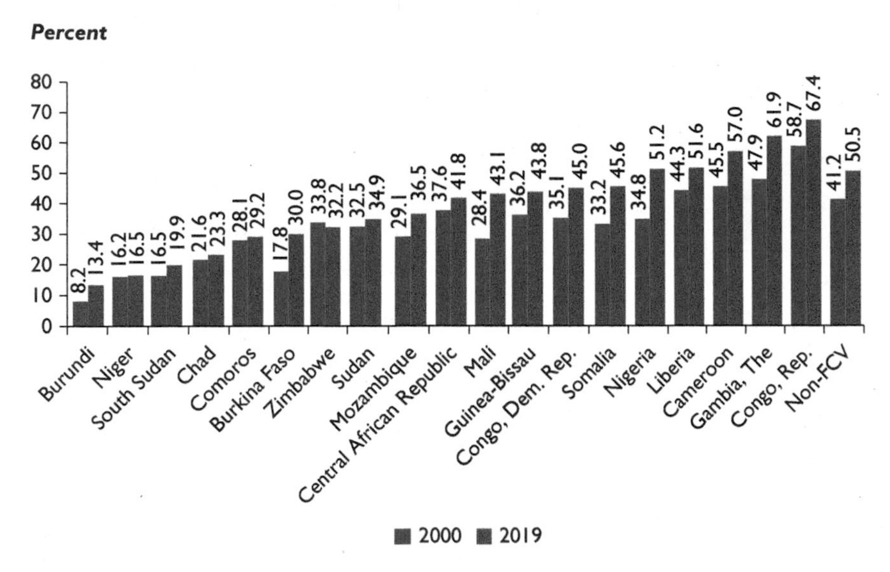

Source: Compilation for this publication, using data from World Bank 2020.
Note: FCV = fragility, conflict, and violence.

share of the population living in urban areas diminished from 33.8 percent in 2000 to 32.2 percent in 2019.

Some African FCV countries had higher outflows of migrants than others. Zimbabwe experienced the most significant net migrant outflow, with nearly 2.5 million people leaving the country from 2002 to 2017 (figure 2.15). Sudan also experienced a large outflow of migrants, losing 2.3 million people from 2002 to 2017. Outmigration from FCV countries became more significant after 2007. Zimbabwe and Sudan lost 1.7 million and 1.8 million people, respectively, while Nigeria lost about 900,000 people and the Central African Republic, Mali, Eritrea, and Burkina Faso all lost more than half a million people in net migration outflows from 2007 to 2017 (World Bank 2020).

Conflict and violence continue to displace people at an increasing rate in Africa. The total number of internally displaced persons (IDPs), refugees, asylum seekers, and stateless individuals in Africa increased from 18.5 million people in 2015 to 33.4 million in 2019 (80.5 percent increase). African FCV countries had to bear the brunt of forced displacement by hosting a large number of refugees. For example, in 2018, Sudan hosted 1.1 million refugees, the Democratic Republic of Congo hosted more than 500,000 refugees, and Chad hosted more than 400,000 refugees (see table 2A.6). Many of these refugees came from other African FCV countries. For example, the number

FIGURE 2.15 Net Migration in African FCV Countries, 2002–17

Five-year estimates, millions

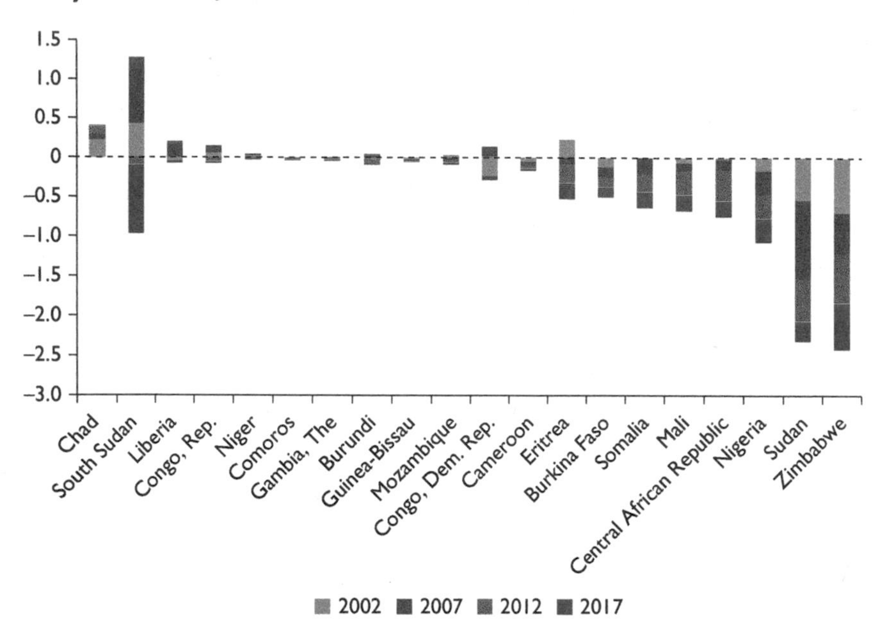

Source: Compilation for this publication, using data from World Bank 2020.
Note: FCV = fragility, conflict, and violence.

of internationally recognized refugees who fled South Sudan in 2019 was 2.2 million people, followed by Somalia (905,000), the Democratic Republic of Congo (807,000), Sudan (735,000), the Central African Republic (610,000), Eritrea (505,000), Burundi (382,000), Nigeria (296,000), and Mali (164,000) (see figure 2.16). These countries also had many IDPs. According to UNHCR (2020), there were 18.5 million IDPs in Africa by the end of 2019 compared with 10.8 million in 2015, a 40.2 percent increase in just four years. Within Africa, FCV countries accounted for the majority of IDPs. In 2019, six African FCV countries had nearly 83 percent of all of Africa's IDPs: 5.5 million in the Democratic Republic of Congo, 2.6 million in Somalia, 2.6 million in Nigeria, 2.1 million in Sudan, 1.4 million in South Sudan, and 969,000 in Cameroon (figure 2.17). The nexus between violent conflict and population displacement is well documented and provides relevant insights about population movements within and across country borders (Braithwaite, Salehyan, and Savun 2019; UNHCR 2019; Moore and Shellman 2007; Davenport, Moore, and Poe 2003). The literature also shows that population displacement can feed a vicious cycle of violence and insecurity because of the diffusion of conflict across borders (Lischer 2006; Onoma 2013).

 **Number of Refugees, by Country of Origin, 2015 and 2019**

Thousands

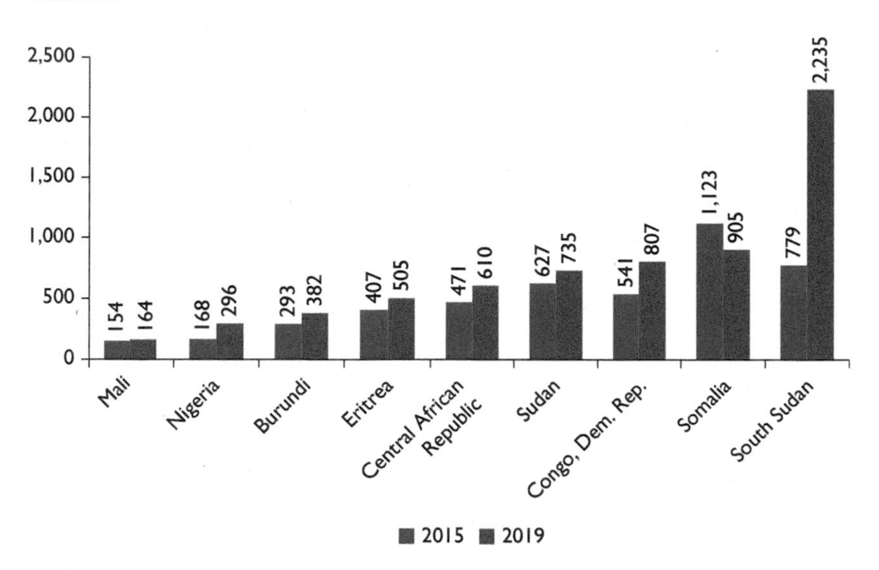

Source: Compilation for this publication, using data from UNHCR 2020.

 **Number of Internally Displaced Persons in 14 African FCV Countries, 2019**

Thousands

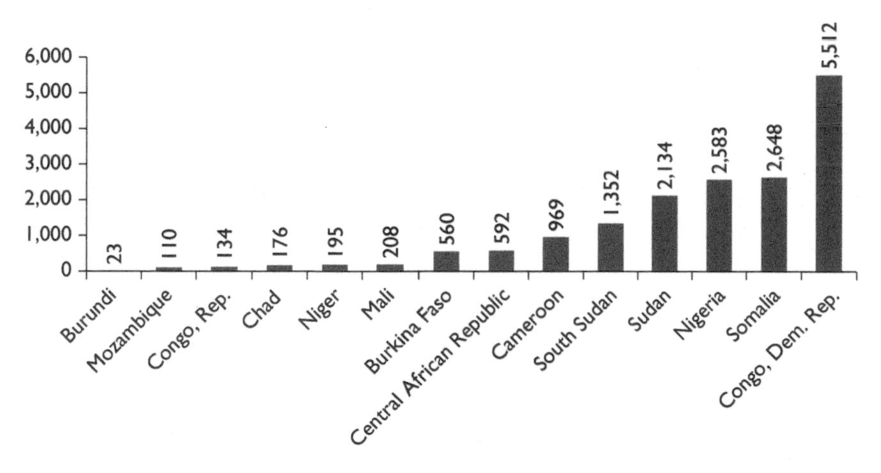

Source: Compilation for this publication, using data from UNHCR 2020.
Note: FCV = fragility, conflict, and violence; IDPs = internally displaced persons.

CLIMATE CHANGE IN FCV COUNTRIES

Climate variability and change will continue to aggravate food insecurity in African FCV countries. A report by the FAO and the United Nations Economic Commission for Africa states that climate change affects African countries "in part because of the heavy reliance on climate-sensitive activities and in part because of the high levels of poverty and food insecurity that exist" (FAO and ECA 2018, 52). The majority of the population in Africa is already experiencing climate variability, which negatively affects food production systems and leads to increased food insecurity (FAO and ECA 2018). Climate change adversely influences crop yields (wheat in particular), fish stocks, and animal health, which reduces overall food supplies and increases food prices, hence inhibiting access to food. Smallholder farmers and poor households are particularly affected by these changes (FAO 2016).

Recurrent droughts and floods affect food security. Droughts reduce the availability of water for crop production and human and animal consumption. As weather patterns become more unpredictable and drought seasons last longer, undernourishment has increased. According to a 2019 report by multiple United Nations agencies (FAO et al. 2019), the number of undernourished people in drought-sensitive countries has increased by 45.6 percent since 2012. During the past decade, droughts have disrupted agricultural activity in several African FCV countries, including Chad, the Democratic Republic of Congo, Somalia, South Sudan, Sudan, and Zimbabwe (FAO and ECA 2018). Drought and conflict affected food availability and prices in South Sudan and Sudan, and climate shocks including droughts disrupted agricultural production and worsened food security in Djibouti, Mozambique, and Somalia (FAO and ECA 2018). Floods also negatively affect agricultural outputs, leading to food shortages in Sub-Saharan Africa (Kotir 2011). For example, in 2020, widespread flooding affected the food security of 4 million people in East Africa (World Vision 2021).

Climate change is causing temperatures to rise and water resources to dwindle in African FCV countries, which hampers these countries' food production systems. Figure 2.18 shows that temperatures have risen over the past two decades in Africa's FCV countries. These increases have been higher than half a degree Celsius in 18 of the 20 African FCV countries. Evidence suggests that rising temperatures threaten wheat and maize production, lower forest productivity, increase fire risks, negatively affect fisheries, and cause imbalances in fragile ecosystems, changing pest and crop disease patterns (FAO and ECA 2018). Consequently, renewable freshwater resources declined in all the African FCV countries from 2002 to 2017. These declines were most dramatic in the Democratic Republic of Congo and Liberia, where each country's per capita renewable freshwater resources declined by more than 23,000 cubic meters (figure 2.19). The depletion of renewable freshwater resources affects food production capabilities, particularly in countries where freshwater withdrawals for agricultural purposes are significant. Climate change is projected

FIGURE 2.18 Change in Average Temperature from 2000 to 2016 in African FCV Countries

Degrees Celsius

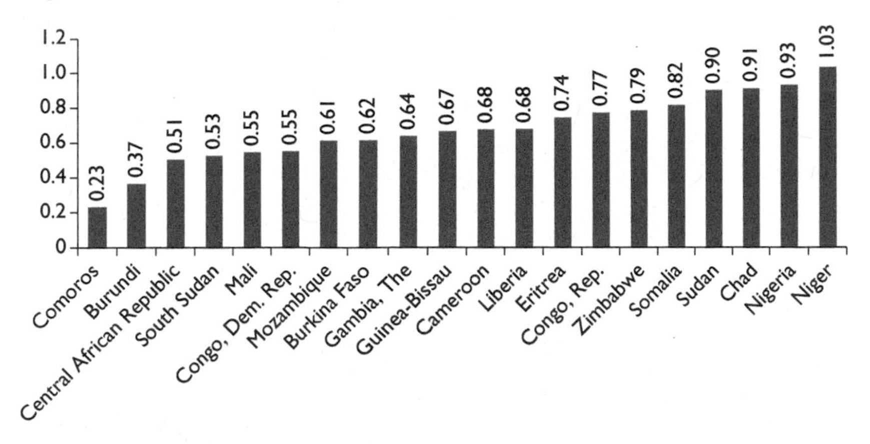

Source: Compilation for this publication, using data from the World Bank Climate Change Knowledge Portal.

Note: FCV = fragility, conflict, and violence.

FIGURE 2.19 Changes in Renewable Freshwater Resources from 2002 to 2017 in African FCV Countries

Cubic meters per capita

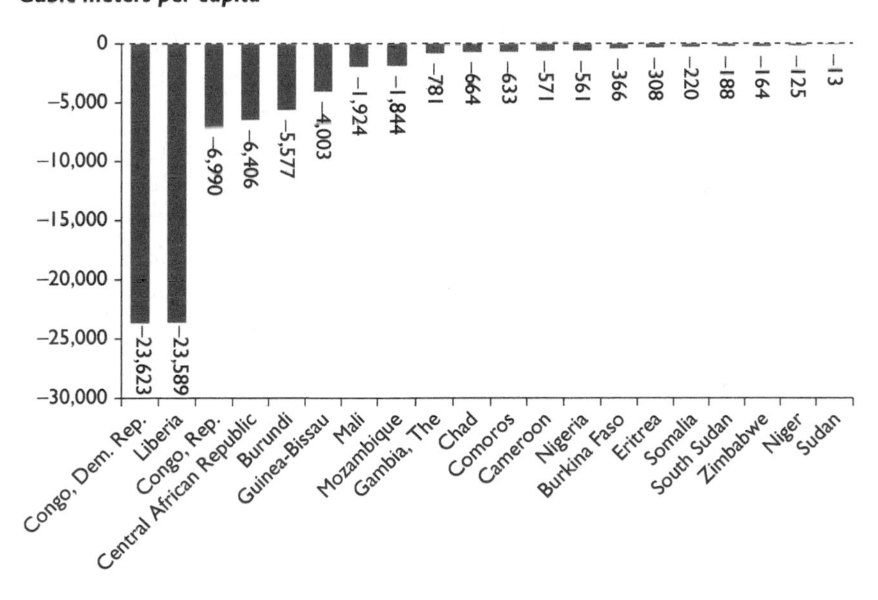

Source: Compilation for this publication, using data from the World Bank Climate Change Knowledge Portal.

Note: For South Sudan and Sudan, the change is from 2012 to 2017. FCV = fragility, conflict, and violence.

to reduce access to water further in African countries, affecting incomes, sanitary conditions, and food availability (FAO and ECA 2018). This is troubling for the agriculture sector, which requires large amounts of freshwater. In 2017, agriculture was responsible for over 70 percent of water withdrawals in 10 African FCV countries: Somalia (99.5 percent), Mali (97.9 percent), Sudan (96.2 percent), Eritrea (94.5 percent), Niger (87.7 percent), Zimbabwe (83.0 percent), Cameroon (79.3 percent), Chad (76.4 percent), Guinea-Bissau (75.8 percent), and Mozambique (73.0 percent). Agriculture was not a significant source of water withdrawals in the Central African Republic (0.6 percent), the Democratic Republic of Congo (4.4 percent), Liberia (8.4 percent), or the Republic of Congo (10.5 percent) (table 2.5).

TABLE 2.5 Annual Freshwater Withdrawals for Agriculture as a Percentage of Total Freshwater Withdrawals, 2002 and 2017

Country	2002	2017
Burkina Faso	55.6	51.4
Burundi	72.6	67.7
Cameroon	77.9	79.3
Central African Republic	0.6	0.6
Chad	78.0	76.4
Comoros	47.0	47.0
Congo, Dem. Rep.	8.7	4.4
Congo, Rep.	11.5	10.5
Eritrea	84.4	94.5
Gambia, The	41.3	38.6
Guinea-Bissau	79.6	75.8
Liberia	9.0	8.4
Mali	98.1	97.9
Mozambique	75.5	73.0
Niger	67.1	87.7
Nigeria	51.0	44.2
Somalia	99.6	99.5
South Sudan	36.5	36.5
Sudan	96.2	96.2
Zimbabwe	78.9	83.0
Average African FCV	58.4	58.6

Source: Compilation for this publication, using data from World Bank 2020.
Note: For South Sudan and Sudan, the values in the first column are for 2012.

ANNEX 2A

TABLE 2A.1 Undernourished People, 2000–18

Millions

Country	2000	2005	2010	2015	2018
Burkina Faso	2.93	3.09	3.01	3.19	3.79
Cameroon	3.68	2.86	1.87	1.56	1.59
Chad	3.38	3.83	4.82	5.00	6.13
Congo, Rep.	0.87	1.24	1.44	1.23	1.47
Gambia, The	0.24	0.34	0.24	0.24	0.27
Liberia	1.08	1.16	1.25	1.74	1.81
Mali	1.85	1.72	1.31	0.94	0.97
Mozambique	6.67	6.84	5.69	8.38	9.62
Nigeria	11.41	10.28	11.73	20.11	24.68
Sudan	6.07	6.59	7.53	4.43	5.18

Source: Compilation for this publication, using data from World Bank 2020.

TABLE 2A.2 Children under Age Five Who Are Stunted

Country	Year	Rate (%)	Children (thousands)
Burkina Faso	2018	24.9	835
Burundi	2019	54.2	1,091
Cameroon	2018	28.9	1,193
Central African Republic	2018	40.8	299
Chad	2015	39.8	1,047
Comoros	2012	31.1	35
Congo, Dem. Rep.	2013	42.7	5,754
Congo, Rep.	2014	21.2	165
Eritrea	2010	52.5	279
Gambia, The	2018	13.6	52
Guinea-Bissau	2014	27.6	77
Liberia	2016	30.1	212
Mali	2018	26.9	932
Mozambique	2015	42.3	1,957
Niger	2018	48.5	2,141
Nigeria	2018	36.8	12,121

(Continued)

TABLE 2A.2 Children under Age Five Who Are Stunted *(Continued)*

Country	Year	Rate (%)	Children (thousands)
Somalia	2009	25.3	565
South Sudan	2010	31.3	500
Sudan	2014	38.2	2,228
Zimbabwe	2019	23.5	504
FCV	2019	34.2	31,987.2

Source: Compilation for this publication, using data from World Bank 2020.
Note: The values are from the most recent data available. FCV = fragility, conflict, and violence.

TABLE 2A.3 Industry (Manufacturing and Construction) Value Added, 2000–19

Percentage of GDP

Country	2000	2005	2010	2015	2019
Burkina Faso	15.4	13.3	12.1	11.6	10.1
Burundi	10.9	11.9	9.2	8.7	—
Cameroon	17.7	15.9	14.5	14.7	14.2
Central African Republic	—	—	18.4	19.7	18.6
Chad	8.6	0.2	1.0	2.8	2.9
Congo, Dem. Rep.	9.9	16.4	16.2	17.1	20.0
Congo, Rep.	3.5	6.4	5.3	7.8	8.6
Eritrea	9.3	6.8	—	—	—
Gambia, The	6.8	6.3	4.6	6.3	4.0
Guinea-Bissau	—	12.8	11.3	10.7	10.5
Liberia	2.5	4.1	2.6	2.1	1.7
Mali	—	7.0	6.7	6.4	6.6
Mozambique	12.8	13.5	10.0	8.2	9.1
Niger	9.2	8.4	6.9	7.6	7.1
Nigeria	13.9	10.1	6.6	9.4	11.5
South Sudan	—	—	2.3	3.5	—
Sudan	8.3	6.7	—	—	—
Zimbabwe	13.4	15.1	9.2	11.9	10.6

Source: Compilation for this publication, using data from World Bank 2020.
Note: GDP = gross domestic product; — = not available.

TABLE 2A.4 Literacy Rate

Percentage of female or male population age 15 years

Country	Year 1	Female	Male	Year 2	Female	Male
Burundi	2000	52.2	67.3	2014	54.7	69.7
Central African Republic	2000	35.3	66.8	2010	24.4	50.7
Chad	2000	12.8	40.8	2016	14.0	31.3
Congo, Dem. Rep.	2001	54.1	80.9	2016	66.5	88.5
Congo, Rep.	—	—	—	2011	72.9	86.4
Comoros	2000	63.5	74.5	2012	42.6	56.5
Côte d'Ivoire	2000	38.6	60.9	2014	36.8	50.7
Eritrea	2002	40.2	65.4	2008	54.8	75.1
Gambia, The	2000	25.1	49.0	2013	33.6	51.4
Guinea-Bissau	2000	27.5	57.6	2014	30.8	62.2
Liberia	—	—	—	2007	27.0	60.8
Mali	2003	15.9	32.7	2015	22.2	45.1
Mozambique	2003	33.2	65.6	2015	43.1	70.8
Sudan	2000	52.1	71.6	2008	46.7	59.8
South Sudan	—	—	—	2008	19.2	34.8
Togo	2000	38.5	68.7	2015	51.2	77.3
Zimbabwe	2011	80.1	87.8	2014	88.3	89.2

Source: Compilation for this publication, using data from World Bank 2020.
Note: The values are from the most recent data available. — = not available.

TABLE 2A.5 Refugees, by Country of Origin, African FCV Countries, 2000–19

Country	2000	2005	2010	2015	2019
Burkina Faso	124	602	1,141	2,147	11,733
Burundi	568,070	438,695	84,053	292,750	381,508
Cameroon	2,067	9,096	14,952	10,565	66,305
Central African Republic	135	42,875	164,902	471,093	610,203
Chad	54,955	48,389	53,713	14,921	11,192
Comoros	25	64	365	550	658
Congo, Dem. Rep.	371,705	430,919	476,691	541,487	807,374

(Continued)

TABLE 2A.5 Refugees, by Country of Origin, African FCV Countries, 2000–19 *(Continued)*

Country	2000	2005	2010	2015	2019
Congo, Rep.	27,570	24,428	20,682	14,772	13,481
Eritrea	376,847	144,062	222,454	407,428	505,118
Gambia, The	751	1,680	2,233	8,473	17,813
Guinea-Bissau	880	1,042	1,117	1,472	2,093
Liberia	266,926	231,136	70,134	9,972	5,375
Mali	364	512	3,659	154,202	164,461
Mozambique	24	93	120	55	77
Niger	483	642	794	1,383	3,063
Nigeria	5,735	22,123	15,645	167,973	295,569
Somalia	475,644	395,544	770,143	1,123,144	905,109
South Sudan	—	—	—	778,722	2,234,814
Sudan	494,351	693,621	387,265	627,080	734,947
Zimbabwe	109	11,246	24,080	21,332	10,603
Total FCV	2,646,765	2,496,769	2,314,143	4,649,521	6,781,496

Source: Compilation for this publication, using data from UNHCR 2020.
Note: FCV = fragility, conflict, and violence; — = not available.

TABLE 2A.6 Refugees, by Asylum Country, 2005–18

Country	2005	2010	2015	2018
Burundi	20,681	29,365	53,363	71,507
Central African Republic	24,569	21,574	7,330	6,655
Chad	275,412	347,939	369,540	451,210
Comoros	1	—	—	—
Congo, Rep.	66,075	133,112	44,955	37,494
Congo, Dem. Rep.	204,341	166,336	383,095	529,061
Côte d'Ivoire	41,627	26,218	1,980	1,810
Djibouti	10,456	15,104	19,365	18,295
Eritrea	4,418	4,809	2,549	2,252
Gambia, The	7,330	8,378	7,854	4,034
Guinea-Bissau	7,616	7,679	8,684	4,850
Liberia	10,168	24,735	36,505	9,122
Mali	11,233	13,558	15,917	26,539

(Continued)

TABLE 2A.6 Refugees, by Asylum Country, 2005–18 *(Continued)*

Country	2005	2010	2015	2018
Mozambique	1,954	4,077	5,622	4,907
South Sudan	—	—	263,016	291,842
Somalia	493	1,937	8,081	16,741
Sudan	147,256	144,008	309,639	1,078,287
Togo	9,287	14,051	21,953	12,336
Zimbabwe	13,850	4,435	6,950	7,797
MENA	—	1,889,712	2,675,408	2,649,792
Africa	—	2,348,368	4,769,513	6,745,759

Source: Compilation for this publication, using data from UNHCR 2020.
Note: MENA = Middle East and North Africa; — = not available.

NOTES

1. The list is available at https://www.worldbank.org/en/topic/fragilityconflictviolence/brief/harmonized-list-of-fragile-situations and was downloaded on March 28, 2021. Libya, which is part of the World Bank's Middle East and North Africa region, is not included in the analysis.
2. As of 2021, 20 countries in the Africa region were classified as FCV, but undernourishment data are available for only 10 of those countries.
3. Malnutrition includes undernutrition (undernourishment, wasting, stunting, and underweight), inadequate vitamins or minerals, excess weight, obesity, and resulting diet-related noncommunicable diseases.
4. ASF includes a wide variety of meat and dairy products, including fish, fowl, milk, eggs, insects, and other small animals.
5. The economic structural transformation of agriculture usually follows a common path. This path starts with mechanization and enhanced farming practices (including the use of fertilizers, pesticides, and tilling practices, among others) increasing agricultural output because of higher labor and land productivity. This then reduces the need for labor and displaces rural workers, which leads to rural-to-urban migration and increased economic activity in urban areas, particularly in the manufacturing and service sectors (World Bank 2018b).

REFERENCES

ACP-EU (African, Caribbean and Pacific–European Union). 2018. "Agriculture as an Engine of Economic Reconstruction and Development in Fragile Economies." ACP- EU, Brussels.

Braithwaite, A., I. Salehyan, and B. Savun. 2019. "Refugees, Forced Migration, and Conflict: Introduction to the Special Issue." *Journal of Peace Research* 56 (1): 5–11.

Brazier, Y. 2019. "What Is Nutritional-Deficiency Anemia?" *MedicalNewsToday*, December 6, 2019. https://www.medicalnewstoday.com/articles/188770.

Davenport, C., W. Moore, and S. Poe. 2003. "Sometimes You Just Have to Leave: Domestic Threats and Refugee Movements, 1964–1989." *International Interactions* 29 (1): 27–55.

FAO (Food and Agriculture Organization of the United Nations). 2008. "An Introduction to the Basic Concepts of Food Security." FAO, Rome. http://www.fao.org/3/al936e/al936e00.pdf.

FAO (Food and Agriculture Organization of the United Nations). 2014. "Promoting Economic Diversification and Decent Rural Employment towards Greater Resilience to Food Price Volatility." FAO, Rome.

FAO (Food and Agriculture Organization of the United Nations). 2016. *The State of Food and Agriculture 2016: Climate Change, Agriculture and Food Security.* Rome: FAO.

FAO (Food and Agriculture Organization of the United Nations). 2018. "Crop Prospects and Food Situation." Quarterly Global Report 2, June, FAO, Rome.

FAO and ECA (Food and Agriculture Organization of the United Nations and European Commission on Agriculture). 2018. *Regional Overview of Food Security and Nutrition: Addressing the Threat from Climate Variability and Extremes for Food Security and Nutrition.* Accra, Ghana: FAO.

FAO, IFAD, UNICEF, WHO, and WFP (Food and Agriculture Organization of the United Nations, International Fund for Agricultural Development, United Nations Children's Fund, World Health Organization, and World Food Programme). 2017. *The State of Food Security and Nutrition in the World 2017: Building Resilience for Peace and Food Security.* Rome: FAO.

FAO, IFAD, UNICEF, WHO, and WFP (Food and Agriculture Organization of the United Nations, International Fund for Agricultural Development, United Nations Children's Fund, World Health Organization, and World Food Programme). 2019. *The State of Food Security and Nutrition in the World 2019: Safeguarding against Economic Slowdowns and Downturns.* Rome: FAO. https://docs.wfp.org/api/documents/WFP-0000106760/download/?_ga=2.96481119.1160401988.1576434696-1905750771.1576434696.

Galasso, E., and A. Wagstaff. 2018. "The Aggregate Income Losses from Childhood Stunting and the Returns to a Nutrition Intervention Aimed at Reducing Stunting." Policy Research Working Paper 8536, World Bank, Washington, DC.

Hoddinott, J., J. A. Maluccio, J. R. Behrman, R. Flores, and R. Martorell. 2008. "Effect of a Nutrition Intervention during Early Childhood on Economic Productivity in Guatemalan Adults." *Lancet* 371: 411–16.

Horton, S., and R. Steckel. 2013. "Global Economic Losses Attributable to Malnutrition 1900–2000 and Projections to 2050." In *The Economics of Human Challenges*, edited by B. Lomborg. Cambridge: Cambridge University Press.

IFPRI (International Food Policy Research Institute). 2015. *The Global Nutrition Report 2015.* Washington, DC: IFPRI. doi:10.2499/9780896298835.

Koren, O., and B. E. Bagozzi. 2017. "Living Off the Land: The Connection between Cropland, Food Security, and Violence against Civilians." *Journal of Peace Research* 54 (3): 351–64.

Kotir, J. 2011. "Climate Change and Variability in Sub-Saharan Africa: A Review of Current and Future Trends and Impacts on Agriculture and Food Security." *Environment, Development and Sustainability* 13: 587–605. 10.1007/s10668-010-9278-0.

Lischer, S. K. 2006. *Dangerous Sanctuaries: Refugee Camps, Civil War, and the Dilemmas of Humanitarian Aid.* Ithaca, NY: Cornell University Press.

Martin-Shields, C., and W. Stojetz. 2018. "Food Security and Conflict: Empirical Challenges and Future Opportunities for Research and Policy Making on Food Security and Conflict." FAO Agricultural Development Economics Working Paper 18-04, Food and Agriculture Organization of the United Nations, Rome.

Moore, W. H., and S. M. Shellman. 2007. "Whither Will They Go? A Global Study of Refugees' Destinations, 1965–1995." *International Studies Quarterly* 51 (4): 811–34.

Onoma, A. K. 2013. *Anti-Refugee Violence and African Politics*. Cambridge: Cambridge University Press.

Population Action International. 2011. "Why Population Matters to Food Security." Population Action International, Washington, DC. https://www.everythingzed.rf.gd/edu/DATA/ETZed_PAI-1293-FOOD_compressed-1.pdf.

Shekar, M., J. Kakietek, J. Dayton Eberwein, and D. Walters. 2017. *An Investment Framework for Nutrition: Reaching the Global Targets for Stunting, Anemia, Breastfeeding, and Wasting*. Directions in Development. Washington, DC: World Bank.

Skoufias, S. 2018. "All Hands on Deck: Halting the Vicious Circle of Stunting in Sub-Saharan Africa." *Investing in Health* (blog), November 28, 2018. https://blogs.worldbank.org/health/all-hands-deck-halting-vicious-circle-stunting-sub-saharan-africa.

Timmer, C. P. 2017. "Food Security, Structural Transformation, Markets and Government Policy." *Asia and the Pacific Policy Studies* 4 (1): 4–19.

UNDP (United Nations Development Programme). 2012. "Food Production and Consumption Trends in Sub-Saharan Africa: Prospects for the Transformation of the Agricultural Sector." Working Paper 2012-011, UNDP, New York. https://www.africa.undp.org/content/dam/rba/docs/Working%20Papers/Food%20Production%20and%20Consumption.pdf.

UNHCR (United Nations High Commissioner for Refugees). 2019. *Global Trends: Forced Displacement in 2018*. Geneva: UNHCR.

UNHCR (United Nations High Commissioner for Refugees). 2020. *Global Trends: Forced Displacement in 2019*. Geneva: UNHCR.

Viteri, F. E. 1994. "The Consequences of Iron Deficiency and Anaemia in Pregnancy on Maternal Health, the Foetus and the Infant." In *Nutrient Regulation during Pregnancy, Lactation, and Infant Growth: Advances in Experimental Medicine and Biology*, vol. 352, edited by L. Allen, J. King, and B. Lönnerdal. Boston, MA: Springer. https://doi.org/10.1007/978-1-4899-2575-6_10.

World Bank. 2006a. *Why Invest in Nutrition?* Washington, DC: World Bank. http://siteresources.worldbank.org/NUTRITION/Resources/281846-1131636806329/NutritionStrategyCh1.pdf.

World Bank. 2006b. *Repositioning Nutrition as Central to Development: A Strategy for Large-Scale Action*. Washington, DC: World Bank.

World Bank. 2017. *An Investment Framework for Nutrition: Reaching the Global Targets for Stunting, Anemia, Breastfeeding, and Wasting*. Directions in Development. Washington, DC: World Bank.

World Bank. 2018a. *Thinking CAP: Supporting Agricultural Jobs and Incomes in the EU*. Washington, DC: World Bank.

World Bank. 2018b. *Exploring the Potential of Agriculture in the Western Balkans*. Washington, DC: World Bank.

World Bank. 2020. *World Development Indicators*. Washington, DC: World Bank.

WHA (World Health Assembly). n.d. "WHA Global Nutrition Targets 2025: Stunting Policy Brief." World Health Organization, Geneva. https://www.who.int/nutrition/topics/globaltargets_stunting_policybrief.pdf.

World Vision. 2021. "Africa Hunger, Famine: Facts, FAQs, and How to Help." World Vision, Federal Way, WA, https://www.worldvision.org/hunger-news-stories/africa-hunger-famine-facts.

CHAPTER THREE

Understanding Insect Farming

HIGHLIGHTS

- Insect farming contributes to food security, job and livelihood creation, and environmental protection, while reducing waste.
- Farmed insects grow faster; require fewer inputs like water, land, and feed; and have lower climate and environmental impacts compared with other livestock species.
- Insect consumption has a long history in Africa, but insect farming is much more recent. Africa consumes about 25 percent of the 2,100 insect species recognized as edible worldwide. Among these, approximately 18 are suitable for farming.
- Insect farming is a rapidly growing industry. Estimates show that the market for insects as food and animal feed will be worth up to US$8 billion by 2030, a 24 percent compound annual growth rate (CAGR) over the decade.
- Africans are willing to farm insects. A study of Ugandan fish farmers shows that over 90 percent were willing to use insects for feed, but fewer than half ever had.
- In Africa, insect farming institutions and regulatory frameworks are still in their infancy.
- Insects provide similar levels of protein and micronutrients as animal products and could functionally replace fish- and soybean-based feeds, among others.
- Insect farming can increase food security and employment opportunities for vulnerable groups like youth, women, and the displaced.

- Replacing livestock rearing with insect farming can drastically reduce the emission of harmful greenhouse gases (GHGs) and limit degradation caused by wild insect harvesting.
- Insects convert organic agricultural and industrial waste into edible protein. This makes insects, particularly the black soldier fly, ideal for breeding in African countries affected by fragility, conflict, and violence (FCV), where organic waste management is a challenge.
- Insect biofertilizer, made from insect farming waste, has a comparable nutrient composition as traditional fertilizers but with a fraction of the energy and fossil fuel requirements.
- The insect farming industry is profitable, including for small-scale farmers. In Cameroon, the average income of formal African palm weevil larvae collectors represented 30 to 75 percent of their household income. In Ghana, palm weevil farmers could pay back their initial capital investment in 127 days.
- The prices for farmed insect products remain high, but evidence suggests that they are becoming more competitive with other protein products, such as soy and fishmeal, as the industry matures.

The purpose of this chapter is to understand the state of insect farming in Africa and its many benefits as a frontier agricultural technology within a circular food economy. The chapter shows that insects can be farmed as micro-livestock, which grows faster, requires fewer inputs like water and feed, and has lower climate and environmental impacts than other livestock species. The first section looks at the context of insect farming in Africa and finds that insect consumption has a long history on the continent, the industry is growing, and farmers seem willing to try it. The second section looks at the types of insects that can be farmed and consumed and finds that there is a wide variety of insects and uses. The third section looks at the roles played by civil society, government, and the private sector in the insect sector. It takes a close look at the Republic of Korea's advanced regulatory framework, as a case study, and finds that regulatory frameworks in Africa are still in their infancy. It also finds that private businesses drive innovation and employment in the sector. The fourth section assesses the nutritional benefits for humans and animals of consuming insects. It finds that insects provide similar levels of protein and micronutrients as animal products and have health benefits for humans and livestock. The fifth section reviews the social benefits from insect farming, finding that it may benefit vulnerable groups like youth, women, and the displaced. The sixth section examines the environmental benefits of insect farming. It finds that replacing livestock rearing with insect farming can drastically reduce the emission of harmful GHGs, and that insect farming can efficiently dispose of human and industrial waste, protect endangered plants and wildlife, and contribute to soil health when insect frass (manure) is used for fertilizer. The seventh section looks at the economic benefits of insect farming, particularly the industry's profitability and prices for insect products. It finds that the industry is profitable, especially for small-scale farmers, and that prices remain high, but these will decline as the industry matures.

CONTEXT OF INSECT FARMING IN AFRICA

History

Insects are part of traditional diets in Africa. More than 400 of the 2,100 insect species recognized as edible worldwide (Jongema 2017), or nearly 25 percent, are consumed in Africa (map 3.1) (Kelemu et al. 2015). The wild harvested insect species consumed in Africa are a diverse set that includes *Blattodea* (cockroaches), *Coleoptera* (beetles and grubs), *Diptera* (flies), *Hemiptera* (true bugs), *Hymenoptera* (ants, bees, and wasps), *Isoptera* (termites), *Lepidoptera* (butterflies and moths), *Odonata* (dragonflies and damselflies), and *Orthoptera* (grasshoppers, crickets, and locusts). Many refugees, such as Central Africans, Congolese, and South Sudanese, have experience collecting insects in their home countries. Congolese and South Sudanese refugees in the Kakuma refugee camp in northwestern Kenya are experimentally farming field crickets (DanChurchAid 2020).

The traditional consumption of wild harvested insects in Africa varies among countries and regions. Consuming insects has historically been a part of the food culture of many African countries affected by FCV (refer to annex 3A).[1] Among African FCV countries, the Central African Republic, the Democratic Republic of Congo, the Republic of Congo, and Zimbabwe are the largest consumers of insects (Niassy and Ekesi 2017). For example, 90 percent of Zimbabwe's population has consumed insects, with termites and mopane caterpillars being the most common (Dube et al. 2013). Insects are generally collected from the wild during peak seasons and consumed at home, with the surplus sold in local markets. The highest diversity of edible species is found in Central Africa (map 3.1). In the Democratic Republic of Congo, more than 85 different insect species are consumed (Kelemu et al. 2015). In other countries, the history of insect consumption varies among ethnic groups. In Kenya, coastal communities historically have consumed little or no insects (Kelemu et al. 2015), while the Luo population in western Kenya, for example, consumes insects such as termites and lake flies. That said, traditional insect harvesting is rapidly declining among younger generations (Ayieko and Oriaro 2008).

African countries rarely have dietary guidelines, especially related to insects. Dietary guidelines provide guidance for healthy diets and food consumption. Although the Food and Agriculture Organization of the United Nations (FAO) launched a knowledge base on healthy dietary guidelines, the actual issuing of guidelines is each country's national responsibility. By 2018, only seven countries in Africa had officially launched dietary guidelines for healthy food consumption (FAO 2018). The guidelines promote diverse diets with a balanced intake of animal products, particularly fish. However, very few mention insects.

MAP 3.1 Diversity and Abundance of Edible Insects in Africa

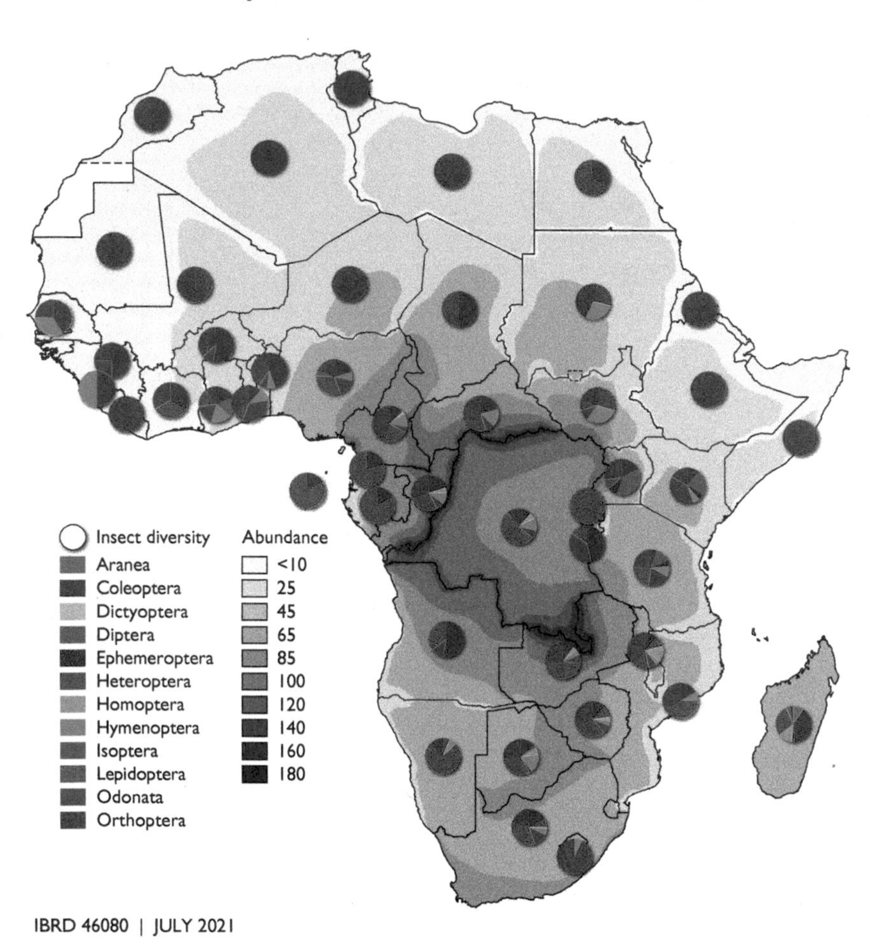

Source: Kelemu et al. 2015.

Insect farming requires little space, unlike other animal production systems. Small-scale insect farms can be housed in homes or small shelters. In Thailand, a cricket powder facility that processed 100 kilograms (kg) of cricket powder per day was only about 60 square meters and required only two workers to operate. Large- and medium-scale insect farming operations require larger warehouse-like facilities. Black soldier fly larvae (BSFL) production facilities are generally located in peri-urban areas, often in industrial parks, close to sources of organic municipal waste, which is BSFL's primary feeding substrate.

Insects have been reared for food, feed, and textile fiber throughout human history. Two insect species in particular have a long history of domestication:

the honeybee (*Apis mellifera)* and several closely related species, and the silkworm (*Bombyx mori*) and a few related moth species (like *Samia ricini*). In the United States, honeybee pollination has an estimated economic value of more than US$20 billion (Degrandi-Hoffman et al. 2019). The rearing of sterile pest insects, such as fruit flies, for integrated pest control is another well-established mass production application (IAEA 2020). In more recent history, simple insect production systems, driven by the need for perennially available animal feed, have been reported since at least the 1960s. Scientific literature shows that people have experimented with domesticating the housefly and black soldier fly (BSF) at small scales for decades and reared mealworms and crickets as live feed for pet animals and zoo reptiles in Western countries (Makkar et al. 2014).

Growing Global Industry

Two surveys that were undertaken for this report shed some light on the scope of insect farming in Africa, but there is still much to be learned. The country-level survey reveals that insect farming occurs in 10 of the 13 surveyed countries on different scales. The country-level survey identified 849 insect farms of various sizes in those 10 countries. However, according to the surveyors, the total number of farms is closer to 1,800. The actual number of insect farms in these countries is unknown beyond these estimates. There are no data on the number of farms in all 54 African countries. The farm-level survey identified 16 farmed or semi-farmed insect species in the 10 countries with insect farming. Most of the farms started after 2010, which shows that insect farming is relatively new in Africa, although some insect farms in Madagascar date back 50 years. Silkworm farming is the oldest form of insect farming in Africa.

Insect farming is a rapidly growing industry. Each year, the number of new entrants, companies, and initiatives increases. The edible insect industry comprises everyone from small-scale farmers and entrepreneurs to large international companies. Barclays Investment Bank estimates that the market for insects as food and animal feed will be worth up to US$8 billion by 2030, a 24 percent CAGR over the decade (MarketWatch 2019). Analysts estimate that the global insect feed market size will increase from US$621.8 million in 2018 to US$1,011.5 million in 2025, a 7.3 percent CAGR (360 Market Updates 2019). The world's compound feed production, or animal feed that is blended from various raw materials, is estimated at over 1 billion tons annually (IFIF 2019). According to the International Feed Industry Federation, global commercial feed manufacturing generates more than US$400 billion in total annual revenue (IFIF 2019). Korea's insect market[2] was valued at ₩264.8 billion (US$220 million) in 2018 and was anticipated to reach ₩361.6 billion (US$290 million) in 2020. Korea's combined sales from insects were ₩37.5 billion (US$31.4 million) in 2018, up 8.7 percent from a year earlier (RDA 2020b). The Thai and Chinese edible insect market demands are projected to surpass US$50 million and US$85 million, respectively, in 2024. At the regional level, the Asia Pacific edible insect market will likely exceed US$270 million by 2024 (Graphical Research 2018). The limited information that exists on the economics of insect farming suggests that the industry will continue to grow and become more profitable.

The insect farming industry is also growing in Europe. The International Platform of Insects for Food and Feed estimates that insect protein production in Europe will exceed 1 million tons by 2025 and 3 million tons by 2030 (IPIFF 2019). This amount will surpass the 0.32 million tons of fishmeal protein currently used in Europe (European Commission 2019). European insect producers have commercialized more than 5,000 tons of insect protein since legislation authorizing insect protein for use in aquaculture feed came into effect in 2017. As of 2019, European insect producers raised more than US$660 million through investments and expected to raise more than US$2.75 billion by the mid-2020s. Directly and indirectly, this increase is expected to create more than 100,000 jobs by 2030 (figure 3.1) (IPIFF 2019). There are no similar statistics for the African continent.

Globally, private sector investment is increasing rapidly in the large-scale production and supply to the marketplace of insect protein. Large companies and suppliers in the agri-food sector—such as the Bühler Group (Buhler Group 2019), Cargill (Byrne 2018a), WIlbur-Ellis (Wilbur-Ellis 2018), and McDonald's (Byrne 2018b)—are increasingly investing in novel insect-based proteins as feed sources. The rapid CAGR in the insect farming sector has attracted investor interest from notable venture capital firms around the world, including Upfront Ventures, Astenor Ventures, and the Footprint Coalition, among others. In Europe, Astenor Ventures invested US$372

FIGURE 3.1 Number of Direct and Indirect Jobs Created in the Insect Food and Feed Industry in Europe

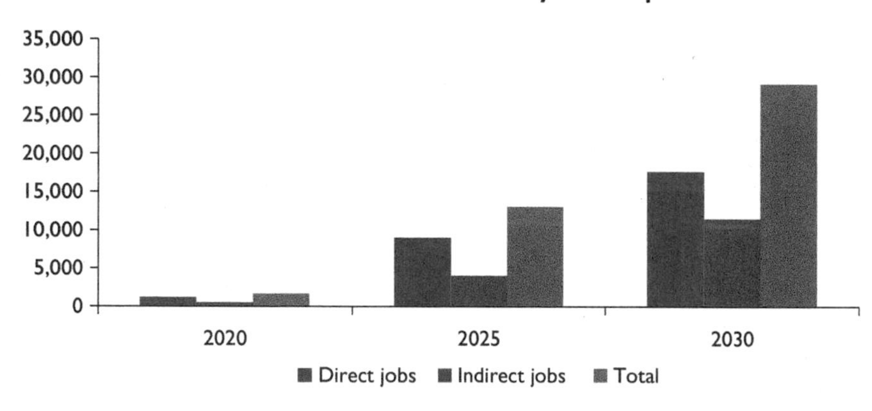

Source: IPIFF 2021.

million in Ynsect, a French insect feed producer (Law 2020; Reuters 2020). In 2020, Rabo Corporate Investments, the financial services company of Dutch multinational banking company Rabobank, made an undisclosed investment in Protiz,[3] a BSFL producer in the Netherlands. This followed an earlier US$50.5 million investment in Protix from Aqua-Spark, Rabobank, and other investors in 2017 (Jackson 2020). In 2020, Honey Capital invested US$1.2 million in a Hong Kong SAR, China–based agritech startup, FlyFarm,[4] in seed funding to develop a pilot BSFL farm in Australia. In 2013, the US-based Aspire invested US$1.0 million in a palm weevil larvae commercial farming operation in Ghana (AgriTech Capital 2019). Other insect protein producers similarly have been able to attract significant investments. For example, South Africa–based AgriProtein raised about US$105 million for BSFL production by 2018 (AgriProtein 2018).

Willingness to Farm Insects

There is growing interest in farming insects. Increasing pressure to find new ways to feed the growing global population has ignited interest in mass-producing farmed insects. In 2013, the FAO began raising awareness of the potential of domesticating insects for food and feed (van Huis et al. 2013). Since then, modern insect farming has continued to expand as a system for mass-producing edible insects for direct human consumption, or as a protein ingredient in animal feed for fish, poultry, or livestock (Makkar et al. 2014; Chaalala, Leplat, and Makkar 2018).

Farmers are increasingly willing to farm insects for feed. A study of Ugandan fish farmers shows that 95 percent of fish farmers and 92 percent of feed traders and processors were willing to use insects for feed, but only 45 percent of fish farmers and 9 percent of feed traders and processors ever had (Ssepuuya et al. 2019). This willingness among farmers, traders, and processors was because they perceived insects as a source of nutrition in fish diets. The more familiar farmers, traders, and processors were with farming insects for fish feed, the more positive were their perceptions of the practice. Another study in Benin found that 82 percent of poultry farmers were willing to pay for fly larvae meal (Pomalégni et al. 2018). The study also found that these farmers were more educated, earned more, had larger chicken farms, and were more financially dependent on poultry farming than farmers who were unwilling to pay for fly larvae meal. On average, farmers were willing to pay approximately US$0.37 per kilogram for insect-based meal.

A person's culture largely determines his or her willingness to use insects for food or feed. Different cultures have different levels of acceptance toward consuming insects. The technical term for eating insects is "entomophagy," although this term is more widely used to deride insect consumption as strange and undesirable (Evans et al. 2015). This hesitancy to produce insects for human consumption was summed up by one insect farmer from Kenya who said, "One of the challenges we face is discouragement by society, especially from non-insect-eating communities." Many observers describe entomophagy as highly exotic despite the historical use of insects in traditional diets for centuries. As a result, insect consumption is uncommon in many modern societies. That said, studies from Kenya show that consumers are willing to pay for insect-based products (Alemu and Olsen 2018). Another study from Benin shows that consumers are willing to pay more for fish that have been fed insect meal than fish that have been fed soybean or fishmeal (Pomalégni et al. 2018). Another study showed that most Ethiopians would eat cookies containing insect components but only 11 percent of Ethiopians would be willing to eat whole insects (Ghosh et al. 2019). A study in Burundi found that many people did not know that crickets are edible. However, when researchers told them that crickets could be eaten, respondents expressed an interest in consuming them. This is because other insect species, like termites and grasshoppers, are a part of Burundi's traditional food culture (TNO 2019). In the same study, urban consumers said they would be willing to pay from US$0.42 in Makamba to US$0.85 in Bujumbura for 250 milliliters of crickets. Another study found that the willingness to taste cricket products varied among nationalities and ethnic groups in the Kakuma refugee camp (table 3.1).

TABLE 3.1 Willingness to Taste a Cricket Product among Nationalities in Kenya's Kakuma Refugee Camp, 2016

Country	Ethnicity	Ratio of previous insect eaters to all respondents	Refusal rate (those unwilling to consume insects)	Comments
South Sudan	Dinka	1:15	53%	Insect consumption was virtually absent among the Dinka.
Sudan	Nuba	10:10	0%	All the Nuba had eaten roasted grasshoppers as part of their diet.
Somalia	Darod, Hawiye, Rahanweyn, and Somali Bantus	0:27	48%	No previous insect consumption was mentioned by any respondent.
Democratic Republic of Congo	Bembe	3:7	0%	Rural respondents said insects were a part of their diets, but no urban respondents said the same. Grasshoppers were the most commonly consumed species.
Kenya	Turkana	6:26	15%	Crickets were eaten by the poorest households, especially during famine periods when no other foods were available.

Source: Naukkarinen 2016.

TYPES OF INSECTS THAT CAN BE FARMED

The number of insects identified as suitable for domestication is increasing. Among thousands of edible insect species, approximately 18 have been identified as suitable for farming and upscaled production for animal feed or direct consumption by humans (table 3.2) (van Huis 2019; Halloran et al. 2018). This section reviews several of these. Large-scale insect farming is still in its infancy and more insect species suitable for farming will likely be identified. Like animal domestication, only insect species with optimal characteristics—such as taste, disease resilience, productivity, manageability, and nutritional composition—are likely to be profitable in farming systems. Moreover, the deliberate domestication of insects can lead to higher growth rates, increased feed efficiency, and increased insect tolerance to human handling and crowded conditions (Lecocq 2019).

The ideal insects for mass production have relatively simple life cycles. Insect farming systems are shaped by the life cycles and biological characteristics of the species being produced. The insect species best suited for breeding and production in closed systems have relatively short life cycles and colonizing behavior that thrives in high densities. The species produced in large quantities for food or feed (table 3.2) also have relatively simple life cycles, compared with the two ancient, domesticated insect species, namely the silk moth (for silk) and the honeybee (for honey and pollination). The silk moth (*Bombyx mori*) feeds only on mulberry leaves, limiting substrate options. In addition, farmers selectively breed silkworms to reduce their wing size so they can no longer fly, making farming simpler. Domesticating honeybees is challenging because of their aggression. Insect production systems are adapted to these biological characteristics.

Insects reach maturity faster than most livestock. According to the farm-level survey, all commonly farmed insect species require fewer than four months to mature from egg to harvestable size. Like livestock, insect growth rates depend on the species, climate, quality of substrates, and form of production, among other factors. Table 3.3 outlines the time it takes for different insect species to reach harvestable size from birth. Houseflies and BSFL have the shortest growth period and can be harvested within one to two weeks—for houseflies within as short as four days and up to 20 production cycles per year. By contrast, house crickets can have the longest growth period of all the insect species surveyed and mopane caterpillars can be harvested only twice per year because of long growth periods and seasonal factors.

The farm-level survey revealed that at least 16 different insect species are farmed in 10 of the 13 surveyed African countries (table 3.4). Crickets, mealworms, BSFL, housefly larvae, palm weevil larvae, and mopane caterpillars are the most commonly farmed insects in the surveyed FCV countries. In the Democratic Republic of Congo, the most commonly farmed insect species is the African palm weevil larvae. Insect farming was not reported in Benin and Burundi, and this information was missing from the Tunisia survey.

TABLE 3.2 Most Commonly Farmed Insect Species

Species utilized in full-grown (adult) stages				
Species	**English name**	**For food**	**For feed**	**Notes**
Crickets Order: Orthoptera				
Acheta domesticus	House cricket	X	X (pets)	Farmed for feed for pet birds and reptiles in many countries, particularly in Southeast Asia, especially Thailand. Emerging consumption by people in Europe, Australia, and North America. Farming has expanded to Africa.
Gryllodus sigillatus	Banded cricket		X	Same as above. This cricket species is resistant to the *Acheta domesticus* Densovirus paralysis virus, which is known to decimate cricket populations.
Gryllus assimilis	Field cricket	X	X	Originates from the Caribbean. Commonly used in the US pet food industry. Resistant to the *Acheta domesticus* Densovirus paralysis virus.
Gryllus bimaculatus	Black cricket or field cricket	X		Widely farmed in Southeast Asia with *Acheta domesticus*. Farming introduced to Europe for food. Resistant to the *Acheta domesticus* Densovirus paralysis virus.
Teleogryllus testaceus (*Gryllus testaceus*)	Common or field cricket	X		Field cricket native to the Americas.
Scapsipedus icipe[a]	Field cricket	X		Field cricket native to East Africa. Recently described as a new species. The species is commonly farmed in Kenya.
Grasshoppers/locusts Order: Orthoptera				
Oxya spp.; *Melanoplus* spp.; *Hieroglyphus* spp.; *Acridia* spp. *Locusta migratora*; *Schistocerca gregaria*	Various species	X	X (pets)	Various grasshopper and locust species are produced for pet food. Some species are marketed for human consumption. In Africa and elsewhere, grasshoppers and locusts are consumed and collected from the wild.

(Continued)

TABLE 3.2 Most Commonly Farmed Insect Species *(Continued)*

Species utilized in larvae stages				
Species	**English name**	**For food**	**For feed**	**Notes**
Mealworms (larvae of darkling beetles) Order: Coleoptera				
Alphitobius diaperinus	Lesser mealworm	X	X	Farmed for human consumption, mainly in Europe and North America.
Tenebrio molitor	Mealworm or yellow mealworm	X	X	Farmed for human consumption, mainly in Europe and North America. Farming of this species for animal feed is expanding. Giant mealworms of this species are treated with juvenile hormone, an insect hormone that delays metamorphosis.
Zophobas morio	Superworms		X (pets)	Commonly farmed for pet reptile food.
Blattidae (*Blattodea*)	Cockroach		X	Cockroaches in China are primarily used for waste management. They are also occasionally farmed for animal feed.
Protaetia brevitarsis	White-spotted flower chafer beetle	X		Chafer beetles are farmed in East Asia (including the Republic of Korea) for consumption and traditional medicine.

(Continued)

TABLE 3.2 Most Commonly Farmed Insect Species *(Continued)*

Other species utilized in pupae or pre-pupae stages				
Species	**English name**	**For food**	**For feed**	**Notes**
Flies Order: Diptera				
Musca domestica	Housefly		X	Housefly larvae are produced on an industrial scale in China and South Africa. Experimental or emerging commercial production takes place in other countries, including in West Africa. Reared for animal feed for fish, pigs, and chicken, as fresh or as a dried and powdered protein supplement.
Hermetia illuscens Harvested as larvae, pupae, or pre-pupae, depending on the production system	Black soldier fly		X	Black soldier fly is the fastest developing insect farming system. There is confirmed industrial production in Europe, Africa, Asia Pacific, and North America. Produced for livestock feed for fish, pigs, and chickens, among others.
Palm weevils Order: Coleoptera				
Rhynchophorus ferrugineus	Red palm weevil	X		Traditionally collected in Southeast Asia. Recently, farming systems have developed in Thailand.
Rhynchophorus phoenicis	African palm weevil	X	X	Traditionally collected in West and Central Africa. Recently, farming systems have developed in Ghana and other countries.
Silkworms Order: Lepidoptera				
Bombyx mori	Silkworm	X	X	A by-product of silk production. Traditionally used for human food and animal feed in Asia. It is used experimentally in processed food and feed products.
Samia ricini	Eri silkworm	X	X	Eri silkworm is a domesticated silk-producing moth that is less commonly farmed because of its less valuable silk. Thailand is experimenting with using Eri silkworm in food products.

Source: Original table for this publication, using observations from various sources, including van Huis et al. 2013; Halloran et al. 2018; and EFSA 2015.

a. Tanga et al. 2018.

TABLE 3.3 Growth Periods and Cycles of the Insect Species Observed in the Farm-Level Survey

Common name	Scientific name	Days to harvest after eggs hatch (range)	Days to harvest after eggs hatch (mean)	Number of annual production cycles (range)	Number of annual production cycles (mean)
Black soldier fly larvae	*Hermetia illucens*	7–50	21	4–12	7
Field crickets	*Gryllus bimaculatus*	40–91	57	2–8	6
House crickets	*Acheta domesticus*	50–119	82	2–5	3
House flies	*Musca domestica*	4–7	14	12–20	14
Mopane caterpillars	*Gonimbrasia belina*	60–90	75	2	2
African palm weevil larvae	*Rhynchophorus phoenicis*	25–60	35	1–12	9
Silkworm crysalids	*Bombyx mori*	35	30–45	1–6	3

Source: Original table for this publication, using the report's farm-level survey in 2019.

TABLE 3.4 Insect Species Farmed for Food and Feed in Africa as Identified in the Farm-Level Survey in 2019

Common name	Scientific name	Uses	Countries where the species is farmed	Number of farms farming the species
Flies	*Hermetia illucens*	Feed, food	Arab Republic of Egypt, Ghana, Kenya, Tanzania, Uganda	14
	Musca domestica	Feed	Ghana	15
Crickets	*Acheta domesticus*	Food, feed	Kenya, Uganda, Zimbabwe	23
	Gryllus bimaculatus	Food, feed	Kenya, Rwanda, Uganda	25
Caterpillars	*Gonimbrasia belina*	Food	Zimbabwe	2
	Cirina butyrospermi	Food	Democratic Republic of Congo	2
	Cymothoe caenis	Food	Democratic Republic of Congo	1
	Imbrasia ertli	Food	Democratic Republic of Congo	1
	Lobobunaea phaedusa	Food	Democratic Republic of Congo	1
African palm weevil larvae	*Rhynchophorus pheoncis*	Food	Cameroon, Ghana	39
Silkworm chrysalis	*Bombyx mori*	Food	Madagascar	33
	Borocera cajani	Food	Madagascar	1
	Samia ricini	Food	Kenya, Madagascar	1
Other species (Nsenene, Eri silkworms, wild silkworms, cockroaches, termites)	*Tenebrio molitor*	Food	Uganda	1
	Periplaneta Americana	Feed	Tanzania	1
	Multiple *Macrotermes* species	Food	Democratic Republic of Congo	1

Source: Original table for this publication, using the report's farm-level survey in 2019.

This includes insects farmed under semi-managed conditions, like caterpillars, where the full insect life cycle is not fully controlled (for more information on insect domestication, see the section in this chapter on insect farming's nutritional benefits). The country-level survey identified a 17th species in Kenya, *Scapsipedus icipe*. Just over three-quarters (76 percent) of the surveyed farms in Africa produce insects for human consumption, while 19 percent of the farms produce insects for fish and livestock feed. Another 5 percent produce insects for both food and feed. Seventy-one percent of the surveyed farms harvest the four most commonly farmed species, namely the African palm weevil larvae, domestic silkworm chrysalis (*Bombyx mori*), and two cricket species (*Gryllus bimaculatus* and *Acheta domesticus*).

ROLES IN INSECT FARMING FOR CIVIL SOCIETY, GOVERNMENT, AND THE PRIVATE SECTOR

Respondents to the country-level survey said various institutions play roles in the edible insect industry. These institutions include the government, private sector, civil society, research organizations, and international agencies (table 3.5). In general, the respondents mentioned more institutions that farm insects for human food than for animal feed. However, respondents also mentioned several institutions that do both. Of the 10 surveyed countries where insects are farmed, 3—Kenya, Madagascar, and Zimbabwe—had institutions that farmed for both the food and animal industries. The institutional mapping confirms that the insect farming industry is multisectoral, linking to health, trade, agriculture, and environmental conservation.

TABLE 3.5 Stakeholders in the Insects as Food and Feed Industry and Their Roles and Functions, 2019

Institutional category	General roles and functions	Insect farming for food	Insect farming for animal feed
Government	• Developing policies • Disseminating information • Providing technical assistance to farmers • Regulating markets • Developing standards	• Ministries of food, agriculture, livestock, or fisheries • Ministry of health • Bureau of standards • National offices of nutrition, nutrition councils, or equivalent • National wildlife authorities • Rural microfinance programs • National agricultural development programs • National boards of trade • Food and drug authorities • Municipal governments	• Ministries of food, agriculture, livestock, or fisheries • Ministry of health • Bureau of standards • Rural microfinance programs • National agricultural development programs • National boards of trade

(Continued)

TABLE 3.5 Stakeholders in the Insects as Food and Feed Industry and Their Roles and Functions, 2019 *(Continued)*

Institutional category	General roles and functions	Insect farming for food	Insect farming for animal feed
Research organizations	• Conducting research • Disseminating information • Providing training to farmers	• Universities • National institutes of agricultural research • International development research institutes • International research projects funded by foreign governments • Entomological societies • International Centre for Insect Physiology and Ecology	• Universities • National institutes of animal or livestock research • International development research institutes • International research projects funded by foreign governments • Entomological societies • International Centre for Insect Physiology and Ecology
Private sector	• Producing, processing, and selling insects • Conducting training and research • Conducting market feasibility assessments • Disseminating information • Providing microloans • Marketing • Advocating for insect farmers	• Hotels • Start-ups and early-stage entrepreneurs • Social enterprises • Food companies • Insect farmers • Insect producers' associations • Savings and credit organizations • Agricultural companies	• Industry boards • Animal feed companies • Fish farming companies • Pet food producers • Insect farmers • Livestock producer associations • Insect producer associations • Agricultural companies
Civil society	• Disseminating information • Providing training and general support to insect farmers	• NGOs • Religious organizations • Foundations • Humanitarian organizations	• NGOs • Religious organizations • Foundations
United Nations and international agencies	• Disseminating information • Providing project funding • Providing microloans	• CGIAR research centers • International development agencies • FAO • IFAD	• Embassies • International development agencies • FAO • IFAD

Source: Original table for this publication, using the report's country-level survey in 2019.
Note: FAO = Food and Agriculture Organization of the United Nations; IFAD = International Fund for Agricultural Development; NGOs = nongovernmental organizations.

Role of Civil Society

The farm-level survey shows that nongovernmental organizations (NGOs) and research institutions carry out most of the insect training and dissemination efforts in Africa. Across Africa, research institutions are conducting collaborative research on insect farming. Known research projects exist in Benin, Burkina Faso, the Democratic Republic of Congo, Ghana, Kenya, Madagascar, Malawi, Nigeria, South Africa, Tanzania, Uganda, and Zimbabwe. Many of these research institutions also train trainers and disseminate insect farming techniques. Individuals and institutions also use social media to disseminate knowledge on insect farming. In Kenya, refugees from Burundi, the Democratic Republic of Congo, and South Sudan are being trained by an NGO in cricket farming and insect processing techniques. The farm-level survey shows that 66 percent of surveyed insect farmers have been trained to farm insects. For the farmers trained, NGOs carried out 87 percent of the training, usually as part of research or development projects. Rarely are these efforts led by public agencies; only 8 percent of farmers have received training from a governmental organization or public extension service, and all but one of those farmers were Malagasy silkworm farmers. Only 2 percent of farmers received training from two or more sources. Ninety percent of trained farmers felt the training was helpful, with 43 percent feeling that the training was "enough to get them established" in insect farming and 47 percent feeling that the training was "sufficient to scale up" their operation. Ten percent of trained farmers said the training was "insufficient."

Networks help to circulate information about insect farming. According to the farm-level survey, personal networks play an important role in disseminating knowledge on insect farming, with 42 percent of farmers receiving their information this way. Only 21 percent of farmers said they seek knowledge on insect farming from the internet. Forty-six percent said they acquire information from two or more information sources. Farmers use popular messaging apps to create groups to share photos, request extra labor during harvests, and solicit advice from other farmers on which traders or wholesalers to trust (Halloran, Roos, and Hanboonsong 2017; Halloran, Roos, Flore, and Hanboonsong 2016). However, the greatest technical knowledge sharing mechanism is farmer-to-farmer interactions. Insect farming requires a lot of trial and error, and many insect farmers and entrepreneurs are self-taught, building the industry from the bottom up. A study in Thailand found that curious farmers visit other farms, even hundreds of kilometers away, to learn how to farm crickets (Halloran, Roos, and Hanboonsong 2017). First-mover farmers often inspire their neighbors to take up cricket farming, and most cricket farmers prefer to have other cricket farmers around them to attract wholesale buyers. A study from Kenya shows that rural farmers are more likely to adopt insect farming if other insect farmers are located nearby (Halloran 2017). A Rwandan farmer described these various means of learning: "In 2016,

I went to DRC and found people eating insects there. They explained to me how insects are nutritious and that [insect farming] can be a business. I started to search on the internet to learn more. I eventually contacted a farmer in the United States who explained how he farms crickets. Later that year I started an insect farm as a trial, and it works well."

Role of Government

This subsection looks at African and non-African governments' roles in promoting and regulating the insect sector. It looks closely at the role the Korean government played in advancing the country's insect farming industry. It finds that Africa's regulatory frameworks are still in their infancy, while Korea's is the most advanced in the world. In general, established agricultural sectors, such as conventional livestock farming, have well-developed regulatory and institutional frameworks. These regulatory frameworks are required to formalize an industry but can act as a barrier to the uptake of disruptive sectors, like insect farming. Some countries and regions develop new regulations to accommodate new industries, while other countries modify existing regulations. This subsection shows examples of both.

Government Regulations

The EU regulations on food safety and animal production affect the edible insect industry (IPIFF 2020). Since 2018, the EU Novel Food regulation has been implemented to authorize insect products for human consumption, requiring that each insect species and product that is marketed for consumers be evaluated by the European Food Safety Authority (EFSA). In 2021, the EFSA published the results from a scientific evaluation of the safety of consuming two insect species: yellow mealworm (*Tenebrio molitor*) and migratory locust (*Locusta migratoria*). The EFSA announced that these two species are safe to consume.[5] As a result, on June 1, 2021, the EU authorized the legal consumption of novel food from yellow mealworm.[6] The EFSA is in the process of evaluating the safety of consuming other insects as well, and the EU is expected to announce more insect species as authorized under the Novel Food regulation for human consumption. Starting in 2017, the EU began implementing regulations authorizing that seven insect species could be used as protein for aquaculture feeds—namely the BSF (*Hermetia illucens*), common housefly (*Musca domestica*), yellow mealworm (*Tenebrio molitor*), lesser mealworm (*Alphitobius diaperinus*), house cricket (*Acheta domesticus*), banded cricket (*Gryllodes sigillatus*), and field cricket (*Gryllus assimilis*). By August 2021, the EU animal feed regulations were further amended to authorize poultry and pigs (porcine animals) to be fed processed insects.[7] At the same time, farmed insects are considered production animals and as such covered by the EU restrictions on the types of substrates given to insects that are intended for animal or human consumption. The following products are prohibited by the EU regulation from being fed to any farmed production animals, including insects: manure, catering waste, slaughterhouse- or rendering-derived products, and unsold meat or fish products from supermarkets or food industries. However, there are some exceptions that allow feed of certain animal origins, such as milk, eggs, honey, rendered fat, and nonruminant animal blood.[8]

Thailand is developing standards for good agricultural practices. The country recently developed a Good Agricultural Practice (GAP) standard for cricket farming. Thailand's National Bureau of Agricultural Commodity and Food Standards under the Ministry of Agriculture and Cooperatives implements the GAP standard. The GAP standard's aim is for farms to produce quality crickets that are safe for consumers, by outlining insect production requirements related to the farm's location, design, layout, administration, disease prevention, and waste management, among others (ACFS 2019). As of March 2020, domestic cricket producers in Thailand were required to have a GAP certification. By following the GAP standard, Thai farmers are prepared for export markets, which have similar quality and safety standards.

Kenya developed a national standard for insect-based animal feed. This national standard (KS 2711:2017) specifies requirements for the dried insect products used as protein for compound animal feeds.[9] Separately, Kenya Wildlife Services requires permits for large-scale insect farming, usually concerning cricket farming. These permits conform to wildlife domestication and livestock transportation regulations. However, there are ongoing discussions among wildlife professionals of whether crickets and other reared insects should be classified as wildlife. The Kenya National Guidelines on Nutrition and HIV/AIDS, for example, recognizes that insects are a part of the country's traditional food culture (Halloran et al. 2015). These Guidelines mention "edible insects such as termites" as potential protein sources along with common animal proteins, such as milk, meat, and fish. The Guidelines recommend that food security in HIV-affected households could be improved by promoting indigenous foods such as termites. Kenya's revised national food composition database includes termites and grasshoppers as consumable foods and outlines their nutrient composition. Despite this progress, many developing countries still do not have standards and regulations governing insects in the food supply chain. This is true in African FCV countries and many others despite those countries having a long history of consuming insects.

The Singapore Food Agency (SFA) authorized using BSFL as fish feed. Specifically, the SFA requires that the substances used to feed the insects be properly overseen and traceable in order for farmers to receive licenses to rear insects. This ensures the safety of the insect-derived animal feed. The SFA also closely monitors insect farming developments, such as new rearing practices, research and scientific literature, and similar regulations adopted by overseas regulatory authorities. The SFA will continue to refine its regulations to support Singapore's insect farming (SG Links 2019).

Korea's Regulatory Framework

Korea has an advanced regulatory and institutional framework for governing the national insect sector. The country's modern insect sector has developed rapidly since 2011. Insect farms are now operating throughout the entire country. According to a Korean government survey, in December 2019, there were 2,535 farms, and by May 2021, there were nearly 3,000 registered farmers producing insects for food and feed in Korea. The emerging sector has created thousands of jobs and incomes for insect farmers and processors (RDA 2020b).

Insect legislation is well established in Korea. The Ministry of Food and Drug Safety approved several insect species as safe for human consumption (2016 Korean Food Standards Codex). These include silkworms and grasshoppers as traditional food ingredients and mealworm larvae, white-spotted flower chafer larvae, dynastid beetle larvae, and two-spotted crickets as new food ingredients. In January 2020, super mealworm larvae (nonfat powder) was temporarily registered as a new food ingredient, and locust and honeybee drone pupae are expected to be registered. Another eight insect species were approved for animal feed through the Feed Management Act. These include dried crickets, dried grasshoppers, mosquito larvae, housefly larvae, mealworms, super mealworm larvae, and BSF larvae and pupae. Insect fat and other approved by-products from insects can also be sold and distributed according to the Enforcement Decree of the Insect Industry Promotion and Support Act.

Insect farming is expanding quickly throughout Korea, and the number of registered insect farms is growing. In 2018, the Ministry of Agriculture, Food and Rural Affairs (MAFRA) registered 2,318 farms and corporations as insect businesses, an 8.5 percent increase from the previous year. The largest proportion were white-spotted flower chafer beetle (*Protaetia brevitarsis*) farms, accounting for 46 percent of all insect rearing farms. The next largest proportions of insect farms were for Rhinoceros beetles (15 percent), crickets (15 percent), mealworms (11 percent), and BSF (2 percent), and other types of insects made up the other 11 percent. Insect farms can be found in all regions, with the highest concentration of farms in the east, south, and northeast of the country. The edible insect market is also growing. Between 2011 and 2018, the insects-for-feed market in Korea grew from US$21 million to US$140 million, and the insects-for-food market grew from US$0 to US$354 million (RDA 2020b). By 2030, Korea's domestic market for insect feed is expected to grow to US$581 million and for insect food to US$815 million.

The central government, mainly MAFRA and the Rural Development Administration (RDA), leads the country's framework, but private businesses and research institutions play a role too. MAFRA plays multiple roles in the nation's insect industry. It developed the country's comprehensive plan for the insect sector and leads insect-related policy development. For example, MAFRA established the Insect Industry Development Support Act. MAFRA also provides pilot grants and financial support to help farmers purchase insect processing machinery. RDA, which is housed within MAFRA, leads the country's insect-related research and development (R&D). RDA's staff investigates, classifies, and identifies food and feed insects. For example, RDA has approved seven insect species for food and feed since 2016. It also carries out research on physio-ecology, pathology, and mass rearing systems for food and feed insects, and it has developed value-added products and industrialization plans. RDA's National Institute of Agricultural Sciences created standard guidelines for insect farming.[10] The government works with private companies and research institutions too. They include the Korean Insect Industry Association, which has 430 members and promotes insect consumption, investigates new market opportunities, publishes knowledge materials, and advocates for the interests of the insect industry. Figure 3.2 maps Korea's national insect framework.

FIGURE 3.2 Korean Government Framework for the Insects-for-Food-and-Feed Industry

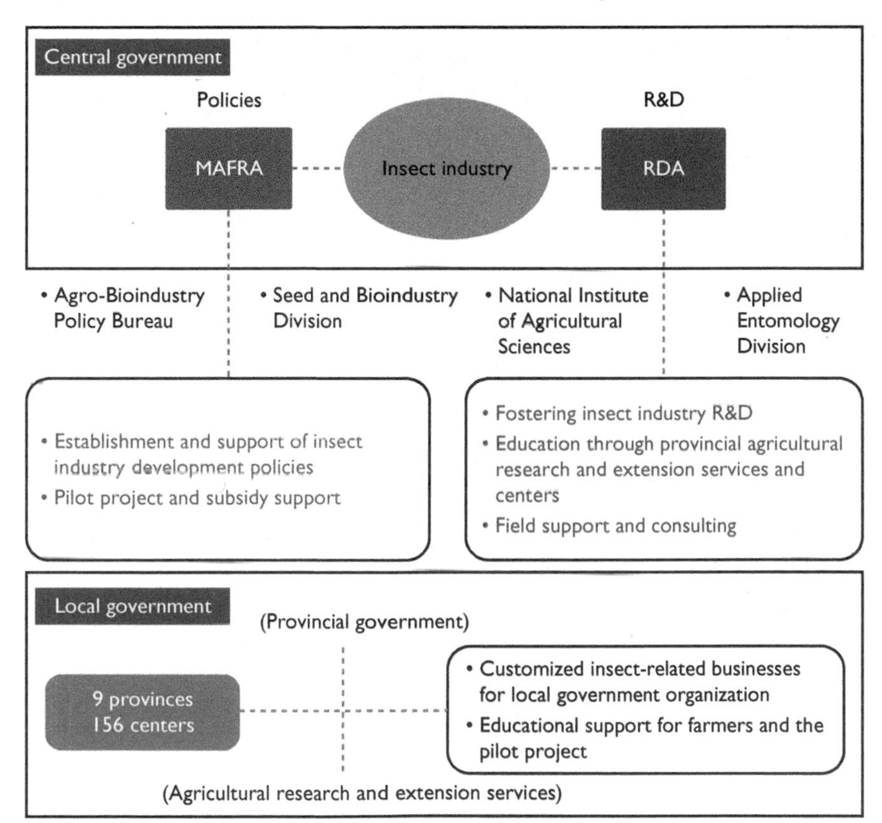

Source: RDA 2020a.

Note: MAFRA = Ministry of Agriculture, Food and Rural Affairs (Republic of Korea); R&D = research and development; RDA = Rural Development Administration (Republic of Korea).

Local governments also support insect farming in Korea. Subnational governments have set up nine agricultural research and extension services and 156 technology centers that carry out trainings and pilot projects for insect farming and related businesses. Local governments also support industrialization by establishing local Insect Resource Centers. These centers finance R&D and insect processing machines, such as microfine grinding mills, and implement educational projects with mentoring and personalized one-on-one training on various insect industry topics. Local governments allow insect farmers to visit the centers and learn by doing by assisting with insect rearing and processing, pest and disease management, and facility and equipment operations. These efforts are done in collaboration with universities and other research institutes. The Insect Resource Centers also promote insect products, and the industry more generally, at exhibitions and media events to raise the public's awareness of the industry's value and benefits.

Korea's Comprehensive Insect Sector Plan, developed by MAFRA, maps out the country's national insect strategy in two phases. The plan describes the status and prospects of the insect industry, a medium- and long-term investment plan, an R&D strategy, the institutional framework of governments and insect-related businesses, and the presidential decree promoting the insect industry. The first five-year phase of the plan lasted from 2011 to 2015, and the second five-year phase of the plan was from 2016 to 2020. The two five-year plans included an investment plan. Between 2011 and 2015, the Korean government allocated US$90 million to develop the insect industry, and between 2016 and 2020, the allocation increased by 20 percent to US$108 million (table 3.6).

TABLE 3.6 Korean Government Areas of Investment for the Country's Insect Sector

First phase five-year plan (2011–15)	Second phase five-year plan (2016–20)
Total budget: US$92.7 million	Total budget: US$117.2 million
• Exploring insect resources and beneficial insects for potential inclusion in the industry: US$4.5 million • Strengthening research and development support to commercialize insect resources: US$14.6 million • Increasing support to insect farming families and for insect industrialization: US$73.6 million	• Advancing consumption and distribution systems: US$3.4 million • Exploring new markets: US$3 million • Building a production base: US$98.2 million • Expanding industrial infrastructure: US$12.7 million

Source: RDA 2020a.

Note: A ₩-USD exchange rate of 0.00084 was used in conversion from Korean won to US dollars.

Phase 1 of the Comprehensive Insect Sector Plan (2011–15) focused on the following:

- *Institutional reform.* Establishing an institutional framework to incorporate the insect industry into the country's existing agricultural paradigm. This included fine-tuning regulations to acknowledge insects as an edible food source for humans and stock feed for animals, establishing a new market for insect products, and adding insects as a temporary food and animal feed ingredient to the official compendium.
- *Production.* Building a robust insect production base. This included establishing four regional insect resource industrialization centers, four insect production complex centers, and four hands-on learning centers, and conducting insect-rearing pilot projects, such as the "Support for the Commercialization of Beneficial Insects Rearing" project to standardize and promote insect rearing, which worked with 39 farming families at eight sites.
- *Research and development.* Financing and implementing 54 research projects.
- *Industrial workforce.* Building professional workforce capacity. This included designating training organizations to develop insect industry experts and providing specialized training to enhance the insect competency of public officials.
- *Promotion.* Increasing the public's general awareness and understanding of the insect industry. This included promoting successful cases of insect production, increasing consumer participation, and holding insect cooking competitions to improve consumer perceptions.

Phase 2 of the Comprehensive Insect Sector Plan (2016–20) focused on the following:

- *Consumption and distribution systems.* Strengthening insect industry systems. This included building insect industry stakeholder networks; enhancing the role of producers and consumers in policy making; enhancing local and regional insect resource industrialization centers; using these centers to connect rearing, distribution, and consumption practices; and carrying out publicity campaigns through diverse media platforms.
- *New markets.* Establishing new markets for insect producers. This included creating a support system for the insect industry, prioritizing insect industries that generate incomes for farming families, finding foreign markets for domestic produce, facilitating entry into those foreign markets, and expanding online and offline channels for insect consumption.
- *Production base.* Building an effective production base. This included increasing the price competitiveness of insect products; increasing the scale and sanitary conditions of insect rearing; financing 120 rearing facilities; establishing an insect management system that matches insect supplies,

particularly silkworm, with demand; preventing the distribution of defective insects; establishing standards for production; strengthening forecasting, diagnoses, and disease control; and exploring a new industrialization model to add value to insect products.
- *Industrial infrastructure.* Expanding the insect sector's industrial infrastructure through R&D investments. This included utilizing industrial technologies, reforming legal and institutional bodies to reduce inefficiencies and advance industrial growth, training a professional workforce, and developing market convergence.
- *Implementation system.* Assigning specific implementation roles and responsibilities for MAFRA, the Ministry of Food and Drug Safety, the Korean Forest Service, local governments, farming families, food producers, the feed industry, distributors, and consumers.
- *Monitoring systems.* Establishing monitoring action plans for RDA, the Korean Forest Service, and local governments and carrying out an annual performance assessment of MAFRA and the comprehensive plan.

Role of the Private Sector

Across the world, the private sector plays an important role in the edible insect industry. Large-scale companies that farm BSF are located in Africa, Asia, Europe, and North America. Most of these companies are rapidly increasing their production volumes, facility sizes, overall investments, and number of employees. Table 3.7 describes some of the large-scale insect farming companies currently operating around the world. Over the past five years, the research team observed more companies operating in the insect farming industry in Africa. Through insect farming, these companies also address societal challenges, such as unemployment, poor sanitation, and gender inequality. For example, one operation, Agriprotein in South Africa, employs more than 150 people in one of Cape Town's most violent and disadvantaged communities, bringing employment and development to the area (Halloran 2018). The following subsection describes examples of successful companies engaged in insect farming in Africa.

Examples of Insect Farming Companies

FasoPro is a social enterprise located in Burkina Faso. The main products it sells are 75 and 150 gram packages of shea caterpillars and 40 gram packages of savory crackers made with shea caterpillar flour. Shea caterpillars are supplied by a network of collectors identified and trained by FasoPro in western Burkina Faso. The company has trained 500 women and collects 15 tons of shea caterpillars each year. As of the first quarter of 2020, the collectors had generated total profits of US$17,000. The shea caterpillars are dried and then transported to Ouagadougou where they are stored. Then, a team ensures the transformation and packaging of the products. Last, a team ensures delivery

TABLE 3.7 Details of Large-Scale Insect Farming Companies, Based on Information Available in 2019

Company name	Insect species	Protein meal production output (tons/year)	Scaled-up capacity target (tons/year)	Oil production (tons/year)	Frass production (tons/year)	Facility location	Employees	Facility size (square meters)	Cost to build (US$, millions)	Annual revenue (US$, millions)	Vertically stacked system?
Agriprotein[a]	Black soldier fly (*Hermetia illucens*)	1,204.50	6,205	3,500	16,500	Cape Town, South Africa Scaled-up locations opening around the world	160	8,500	42	13-15	Missing data
Ynsect[b]	Mealworm (*Tenebrio molitor*)	"Several hundred"	20,000 (at new facility in 2021)	Missing data	Missing data	Burgundy, France New facility being built in Northern France	105	Missing data	Missing data	Missing data	Yes
C.I.E.F.[c]	Black soldier fly (*Hermetia illucens*)	1,095	Missing data	Missing data	Missing data	Jeonbuk province, Republic of Korea	50	Missing data	30	Missing data	Yes
Enterra[d]	Black soldier fly (*Hermetia illucens*)	2,555	Missing data	2,555	2.920	Langley, Canada	32	5,600 (in 2020) 17,000 (future)	30 (future facility)	Missing data	Yes

Source: Original table for this publication.

a. Food Business Africa 2019.

b. Ynsect 2019.

c. Personal communication with the company, 2019.

d. Joly and Nikiema 2019.

of the products to the various points of sale in the country. FasoPro's products are sold in 24 towns at 360 points of sale to thousands of customers. FasoPro pays close attention to the quality of its products, starting from the collection of the raw material, during which specialists supervise and monitor the quality of the caterpillars collected. The collectors are trained, and the raw material is carefully selected to develop the best possible products. The enterprise carries out its processing and production according to good hygiene practices of the food industry. Before packaging and distribution, the products are subject to controls by the National Public Health Laboratory.

Aspire Ghana commercially farms palm weevil larvae and runs a program that empowers peri-rural farmers to raise palm weevil larvae locally. The company employs 60 people and produces 1 ton of palm weevils per month. The containers are stacked. The company focuses on both research and technology development of the farming system as well as commercialization of the larvae.

InsectiPro is an 8,500 square meter BSF farm in Kenya that was producing 3 tons of larvae (wet weight) per day and aimed to increase to 12 tons by the end of 2020. The company employs 46 people and spends US$15,000 per month on labor. Four of the staff members were trained at the International Center for Insect Pathology and Ecology, and two have a degree in animal sciences. InsectiPro also employs an industrial engineer, a food innovation specialist, a human resources officer, a business development consultant, and a BSF specialist. During interviews with the research team, the company's chief executive officer said that she started with very minimal knowledge about how to raise BSF, but she found that most skills can be learned and the trial-and-error method has worked best to innovate and sustain growth. The company plans to sell 80 percent of its product to feed millers and 20 percent directly to farmers. InsectiPro is also developing an outgrower model for smallholder farmers to grow their own BSF and for middle-grower farmers to bring their BSF to InsectiPro to be dried. The project will include 10 local feed millers and approximately 250 farmers to test BSFL in InsectiPro's facilities. InsectiPro initially invested less than US$5,000. To move production up to 1 ton per day, the company invested approximately US$50,000 and then another US$150,000 to take production to its 2020 levels (3 tons per day).

Biobuu Limited currently has factories in Dar es Salaam, Tanzania, and Mombasa, Kenya, and is developing additional factories in Malawi, Uganda, and Zambia. In 2019 interviews, the company said that each factory will employ 20 people directly, taking in 20 tons of waste per day and supplying feed and organic compost to more than 2,000 small- and medium-scale farmers. Biobuu's factory can produce between 1 and 2 tons of insect-derived protein and 5 tons of organic compost per day. The company sells the insect-derived protein as a replacement for soy or fishmeal to local chicken and fish farmers and the compost to local crop farmers. Biobuu says that 1 ton per day is the minimum production needed to provide a decent return on investment. Some of the highest costs are associated with collecting, accessing, and sorting the right type of waste. However, according to the company, municipal governments are often keen to offer land to treat nearby waste.

Sanergy is a Kenyan sanitation company that franchises separation toilets to entrepreneurs in slum areas. The human feces are separated from the urine, and the feces are then used as a substrate to grow BSFL. The BSFL are boiled afterward to kill any pathogens. The larvae are sold to animal feed millers, who grind them into powder mixed with other ingredients. In interviews in 2019, Sanergy said it would open a new facility in Kenya in 2020 that would generate 400 tons of fertilizer and increase BSFL production from 7 to 300 tons per month (Holland 2019).

BioCycle is a community-based production site to treat human feces from everything from individual toilets to large sewer systems. The toilet resources are treated with BSFL. BSFL are the primary agent for turning toilet resources into multiple high-value components: oil (for lubrication and other fuel types), chitosan products, and nitrogen-rich soil conditioners. BioCycle was established in 2013 at the informal settlement of Klipheuwel in Cape Town, South Africa. In collaboration with Ethekwini Water and Sanitation, BioCycle runs a commercial-scale pilot plant with the capacity to process 30 tons of feces from urine-diverting toilets per day. BioCycle is also engineering a bespoke fecal reference plant that will produce data on combining different hazardous resource streams. The initiative will roll out micro businesses to employ local community members, while also improving and increasing local access to sanitation.

INSECT FARMING'S NUTRITIONAL BENEFITS

This section assesses the nutritional benefits for humans and livestock of consuming insects. It finds that insects provide similar levels of protein and micronutrients as animal source foods (ASF) and have health benefits for humans and animals alike.

Edible insects provide protein, a fundamental component for all biological systems. Protein plays an important function in human diets and overall health. Proteins are in the immune system's antibodies, are in the enzymes that drive metabolic functions, and are the core structure of muscle tissue. Protein is made up of 20 organic, nitrogen-containing amino acids: protein's "building blocks." Humans can biologically synthesize 11 amino acids, while human diets must provide the other nine. Along with other ASF—like meat, fish, eggs, and dairy—insects are sources of high-quality protein because they provide high amounts of digestible essential amino acids. The protein quality of insects is considered very good at providing the essential amino acids for human nutrition (table 3.8) (Rumpold and Schluter 2013; Osimani et al. 2018). These amino acids include lysine, leucine, valine, histidine, tryptophan, threonine, methionine, isoleucine, and phenylalanine (FAO and WHO 2007). Studies show that insect protein is also highly digestible, making insects an even more valuable protein source (Poelaert et al. 2018; Longvah, Mangthya, and Ramulu 2011; Jensen et al. 2019).

TABLE 3.8 Fat and Protein in Various Edible Insect Species

Insect species	Common name	Life stage used	Protein (% dry matter)	Fat (% dry matter)
Acheta domesticus	House cricket	Adult	60–75	7–20
Gryllodes sigillatus	Banded cricket	Adult	60–75	7–20
Locusta migratoria	Migratory locust	Adult	40–60	10–25
Hermetia illucens	Black soldier fly	Larvae/pre-pupae	30–60	20–40
Tenebrio molitor	Common mealworm	Larvae	45–60	25–35
Alphitobus diaperinus	Lesser mealworm	Larvae	45–60	25–30
Bombyx mori	Silkworm	Larvae/pupae	50–70	8–10

Sources: Original table for this publication, using values averaged from various sources, including Rumpold and Schluter 2013; Jensen et al. 2019; Beniers and Graham 2019; Irungu et al. 2018.

Researchers are continuously discovering other applications for insect protein. For example, in Korea, insect powder from mealworms (*Tenebrio molitor*) has been tested in hospitals as a protein supplement to help patients, especially elderly patients, recover from various maladies (see photo 3.1). As a result, Korea's MAFRA actively supports developing insect-based foods for health purposes (see photo 3.1 for examples).

Insects are a source of essential nutrients. Insects provide fats and important micronutrients, especially iron and zinc, which are often deficient in food-insecure populations (Black et al. 2013). Protein and fat contents vary among edible insect species depending on the insect's type and development stage (Rumpold and Schluter 2013; Roos 2018). An insect's fat content is specific to that insect's stage of development (examples in table 3.8). These micronutrients are an important contribution to diets in Africa where these minerals are often deficient among children (Black et al. 2013; Holtz et al. 2015). Minerals from insects and animals are characterized by high iron bioavailability (Hallberg et al. 2003) and, therefore, are important in diets dominated by staple plant foods. Iron in edible insect species has been shown to be highly bioavailable in laboratory studies (Latunde-Dada, Yang, and Vera 2016). Consuming the exoskeletons of insects provides chitin, an indigestible fiber. Insect chitin may have probiotic properties that enhance healthy bacteria in digestive systems (Selenius et al. 2018; Stull et al. 2018). One study shows that adding 5 grams of dry insect protein per day to a person's total nutrient intake could alleviate that person's risk of nutritional deficiency of zinc, protein, folate, and vitamin B12 in Africa (Smith et al. 2021).

ASF are important for combatting undernutrition. ASF include all foods that derive from animals, including fish, meat, dairy, and even insects, among many more. In food-insecure situations, households prioritize carbohydrate-rich staple foods to avoid hunger and meet dietary energy needs (Fraval et al. 2019).

PHOTO 3.1 Insect-Based Health Supplements from the Republic of Korea

A mealworm powder product. The label's translation: "Recommended for convalescent patients. Insect processed food. Protein-rich powder"	A mealworm oil capsule. The label's translation: "100% Mealworm. Patent no. 1-1859174. Ministry of Agriculture, Food and Rural Affairs"	A "white-spotted flower chafer beetle" powder. The label's translation: "Fill the man's pride"

But studies show that consuming a diverse diet protects against malnutrition (Development Initiatives 2018). As such, even small amounts of ASF in the diet benefit the nutrition and development of the most vulnerable, particularly children (Skau et al. 2015; Dror and Allen 2011; Bhutta et al. 2008). ASF improve the quality of protein intake and enhance the bioavailability of critical micronutrients. Meat is an important ASF, but unlike insects, red and processed meat is associated with the increased risk of some noncommunicable diseases (Godfray et al. 2018). In high-income countries with industrialized food systems, 60 percent of the protein in diets derives from ASF, in contrast to 20 percent in diets in low-income African countries (IFPRI 2015).

The demand for animal feed has increased in Africa. The rising demand in Africa for meat from fish, poultry, and livestock stimulated the demand for animal feed by 30 percent from 2014 to 2019 (Vernooij and Veldkamp 2019). The demand for meat, milk, fish, and eggs is concentrated in the East and West African countries with the fastest economic growth (Robinson and Pozzi 2011). Kenya, for example, has seen a steep increase in demand for animal feed because of the country's growing demand for animal meat. From 2008 to 2018, Kenya's animal feed production increased from 375,000 to 900,000 tons (Vernooij and Veldkamp 2019).

In 2019, the EAT-Lancet Commission on Healthy Diets from Sustainable Food Systems developed global principles for healthy and sustainable diets (Willett et al. 2019). The Commission's guidelines point at reducing high-income countries' meat consumption. However, the Commission specifically recognized that combatting undernutrition in food-insecure African populations depends on increasing the intake of ASF. In this view, the emerging opportunity to scale up insect farming to meet dietary needs and reduce undernutrition is promising. However, farming edible insects as an alternative to traditional livestock was not included in the global guidelines, which cited the lack of evidence and documentation on insects' health impacts and the benefits of scaling production (Willett et al. 2019).

There is also limited but promising evidence that insects as livestock feed improve animal health and nutrition. The partial or complete replacement of soybean meal or fishmeal in animal feed with insects can provide valuable nutrients and compounds that improve animal microbiota and optimize animal health. While few studies have been conducted on the effects of insect-derived compounds in animal feeding trials, initial investigations are showing great promise. Most of the trial studies that investigate insects as a feed source in animal diets focus on the animal's growth performance, the health and microbiological implications for the animal, and the insect's nutrient composition (Sogari et al. 2019).

Mealworms are a promising aquafeed ingredient. A study tested the effect of yellow mealworm larvae (*Tenebrio molitor*) protein on the growth performance of shrimp (*Macrobrachium rosenbergii*) over 10 weeks (Motte et al. 2019). The shrimp gained the most weight and had the best feed conversion ratio (FCR) when mealworm constituted 50 percent of the shrimp's aquafeed. The study also showed that mealworm feed can be a competitive alternative to traditional aquafeeds (Feng et al. 2019). In another study investigating the digestibility of five different insect meals in Nile tilapia fingerlings, mealworm larvae meal showed the best digestibility, indicating a potential alternative feed for fingerlings (Fontes et al. 2019).

There are opportunities to replace fishmeal with BSFL meal. BSFL meal has been explored as a replacement for fishmeal in trout and salmon diets (Renna et al. 2017) and for African catfish soybean meal diets (Aniebo, Erondu, and Owen 2009). Generally, BSFL meal can replace other protein sources with no negative effects on the fish's growth and survival rate. One study shows that replacing 10–30 percent of fishmeal with BSFL meal in rainbow trout diets modifies the trout's gut microbiota, hence improving the trout's gut health. Compared with fish fed only fishmeal, the insect-based diets induced higher bacterial diversity and more mycoplasma in the fish. These changes in microbiota are attributed to the prebiotic properties of the BSFL's chitin (Terova et al. 2019). The only downside is that replacing fishmeal with BSFL meal reduces the fat quality of certain fish, like trout, because BSFL lacks the healthy long-chained omega-3 fatty acids (docosahexaenoic acid and eicosapentaenoic acid). That said, feeding fish offal to BSFL improves BSFL's fatty acid profile.

Insect meal can substitute for poultry feed ingredients. BSFL can replace fishmeal in chicken feed without any negative effects on the chicken's growth performance (Awoniyi, Aletor, and Aina 2003). Other studies have shown that feeding laying hens BSFL meal did not change the hens' feed intake, body weight, or laying performance (Heuel et al. 2019; Osongo et al. 2018). A meta-analysis of 41 scientific publications that studied the growth performance effects of feeding insects to poultry concluded that partially substituting conventional protein sources with insect species, except grasshoppers, did not diminish the poultry's growth (Moula and Detilleux 2019).

Insect feeds can have antibacterial effects on pig production. In pig feed, BSFL meal can functionally replace other protein sources. Moreover, BSFL can have additional antibacterial benefits for piglets that experience intestinal instability during their weaning period, when solid foods are introduced in their diet. More specifically, the chitin in the BSFL exoskeleton protects the weaning piglets against diarrhea (Ji et al. 2016). BSFL were also found to improve the growth performance of finishing pigs, or fully grown pigs that are being fattened for market (Yu et al. 2019).

INSECT FARMING'S SOCIAL BENEFITS

Insect farming is not associated with any particular gender, but it could provide benefits to women. Historically, wild insect collection in Africa has been a female occupation. A study in Kenya showed that most cricket farmers in Kenya were women (Halloran 2017). In Thailand, in an unrepresentative sample, about half of the cricket farmers who were interviewed for this report were women (Halloran, Roos, and Hanboonsong 2017). That said, anecdotal and observational evidence that was gathered for this report shows that BSF farming is a male-dominated field. The farm-level survey shows that only 28 percent of the surveyed insect farmers were female. The country-level data show that men are overrepresented in BSF and palm weevil farming—86 percent of BSF farmers and 90 percent of palm weevil farmers are male. Despite the lower number of female insect farmers, insect farming still provides an opportunity to empower rural women by increasing their access to livelihoods and agricultural resources. In interviews for this report in Thailand, there was general agreement that cricket farming could help women achieve greater financial independence. However, there is still only limited information available on the relationship between gender and insect farming.

Insect farming benefits people of all ages and income levels. A Kenyan farmer said, "A benefit of insect farming is that it's a source of employment and income for both young and old." The farm-level survey shows that all age groups, including youth, perform insect farming. One percent of insect farmers are between ages 10 and 19 years, 11 percent between 20 and 29, 24 percent between 30 and 39, 20 percent between 40 and 49, 22 percent between 50 and 59, 16 percent between 60 and 69, and 6 percent older than 70. The 40-to-49 age group comprised the most female farmers. Moreover, insect farming attracts people from different economic backgrounds. The country-level survey indicates that 33 percent of the surveyed farmers were identified by the surveyor as "poor," 47 percent were identified as "lower-middle class," 17 percent were identified as "upper-middle class," and 3 percent were identified as "wealthy."

Insect farming can benefit vulnerable populations. The potential of insect farming for vulnerable communities was demonstrated in Kenya's Kakuma refugee camp. It was in this camp that an NGO, in collaboration with Kenya's Jomo Kenyatta University of Agriculture and Technology and financed by Danish Church Aid, trained refugees in cricket farming. The project started in 2017 with a pilot insect farm and training for 15 refugee household heads. By 2021, the project had trained more than 80 household heads in insect farming and processing techniques. These household heads—who have fled a diverse set of countries, including Burundi, the Democratic Republic of Congo, and South Sudan—are now able to produce crickets for animal feed and human food, including for their own consumption. Danish Church Aid plans to scale up the initiative by training more farmers and distributing cricket rearing starter kits to more refugee households. This project shows the potential for insect farming to provide livelihoods and incomes for marginalized communities.

INSECT FARMING'S ENVIRONMENTAL BENEFITS

This section examines the environmental benefits of insect farming. It finds that replacing livestock rearing with insect farming can drastically reduce the emission of harmful GHGs, and that insect farming can efficiently dispose of human and industrial waste, protect endangered plant and wildlife, and contribute to soil health when insect frass is used as fertilizer.

Current food systems contribute to environmental degradation and climate change. There is a growing global recognition that food systems must be reformed to solve the climate and environmental crisis. This is because agriculture has detrimental impacts on natural environments and contributes to climate change. Emerging analyses of how healthy and sustainable diets are obtained support this. Over the past decade, the FAO has focused on how national dietary recommendations can account for GHG emissions, land and water use, and other environmental impacts (Fischer and Garnett 2016).

Insect farming produces fewer GHG emissions and uses less water than traditional livestock production. Insects produce high-quality animal protein with up to 20 times fewer GHG emissions than ruminant livestock, and up to half the emissions of poultry production per kilogram of edible protein (Halloran, Roos, Eilenberg, et al. 2016; Smetana et al. 2016; Halloran 2017; van Huis and Oonincx 2017; Oonincx and de Boer 2012). As such, insect farming could shape future sustainable food systems and contribute to achieving the United Nations Sustainable Development Goals (box 3.1). Farmed insect species also require little water compared with livestock (Miglietta et al. 2015). These insects are produced in high densities, thereby requiring little space and leaving less land vulnerable to exploitation. This is because insects thrive in colonies with very high population densities. In the face of the world's increased demand for protein combined with livestock farming's high GHG emissions, insects represent a promising alternative animal protein source for human food and animal feed (van Huis et al. 2013; van Huis 2019).

BOX 3.1 Insect Farming and the Sustainable Development Goals

Insect farming can address many of the United Nations Sustainable Development Goals (SDGs) by doing the following:

- Supporting livelihood and income diversification—SDG 1
- Providing access to high-quality animal source foods (ASF)—SDG 2
- Providing new livelihood opportunities for men and women—SDG 5
- Reducing waste streams through bioconversion—SDG 6
- Creating jobs in the agriculture and agricultural food sectors—SDG 8
- Fostering innovation and developing a sustainable new industry—SDG 9
- Converting low-grade waste streams into high-quality ASF—SDG 12
- Providing high-quality ASF that emit fewer greenhouse gas emissions—SDG 13
- Providing an alternative to soybean meal and fishmeal in animal feed—SDG 14
- Helping conserve wild insect populations—SDG 15
- Developing multistakeholder partnerships in a new sector—SDG 17

Emissions

Insect farming can reduce GHG emissions. A study in Indonesia found that composting segregated kitchen waste with BSFL can reduce direct carbon dioxide equivalent (CO_2-eq) emissions by 47 times and reduce global warming potential (GWP) by half (Mertenat, Diener, and Zurbrügg 2019). Substituting fishmeal with BSFL meal in animal feed can reduce the GWP by up to 30 percent (Mertenat, Diener, and Zurbrügg 2019). Another study showed that using BSFL meal instead of soybean meal in pig feeds reduced GWP by 10 percent and required 56 percent less land (van Zanten et al. 2018).

Cricket farming also produces lower GHG emissions compared with traditional livestock. A life-cycle assessment of a fully commercialized cricket farming system in Thailand documented that cricket production had 1.5 times fewer GHG emissions than broiler chicken production (figure 3.3), which already has low emission rates compared with ruminant livestock, like cattle (Halloran et al. 2017). The study also found that scaling up cricket farming systems and feeding the crickets efficiently would reduce the system's overall environmental impacts. The main contributors to GWP from insect production were the energy use from heating the insect farms and growing crops to produce insect substrates, such as maize or soybeans (Joensuu and Silvenius 2017; Arru et al. 2019).

Another study shows that insect consumption would reduce GHG emissions. The 2019 study used a country-specific model for healthy, sustainable diets, based on food supply data available from the FAO food statistics database (Kim et al. 2019). In this model, food supply data for 140 countries and nine different scenarios of modified diets for the entire populations were modeled by maintaining a sufficient supply of protein, micronutrients,

FIGURE 3.3 Results of a Life-Cycle Assessment of the Climate Impacts from Farming Crickets, Producing Broiler Chickens, and Optimizing Cricket Farms in Thailand

Global warming potential in kg CO_2-eq

Source: Halloran et al. 2017. Photographs (left and center) © Afton Halloran. Used with the permission of Afton Halloran. Further permission required for reuse; (right) © Nanna Roos / University of Copenhagen. Used with permission from Nanna Roos. Further permission required for reuse.
Note: The "future scenario" for cricket farming was modeled in 2016 for a hypothetical vertical farm. The photo shows a similar vertical system in operation in 2019 at Smile Cricket Farm in Ratchaburi, Thailand. CO_2-eq = carbon dioxide equivalent; kg = kilogram.

and energy. The dietary scenarios ranged from "meatless day" and "no dairy" to scenarios that are partially or fully free of ASF. The modeling included the introduction of edible insects as an option in one scenario called the "low food chain" choice. This scenario included only ASF with low climate impacts and replaced 10 percent of the traditional ASF livestock sources with insects and the other 90 percent with low food chain aquatic sources, such as mollusks and pelagic fish (Kim et al. 2019). The scenario analysis showed that the low food chain scenario, which included edible insects, on an average basis across the 140 countries would reduce GHG emissions by more than 70 percent, the same as the fully ASF-free diet scenario. The low food chain diet was shown to be nutritionally advantageous to a fully vegan diet because the low food chain diet provides all the nutrients required for a healthy diet, including vitamin B12 (cobalamin), which is unlikely to be sufficient in animal-free diets (Wantanabe and Bito 2018).

The main sources of protein in livestock feed come from soybean and fishmeal. Limited natural resources—primarily land—and the increasing demand for animal protein contribute to the demand for livestock feed. Soybean is a preferred ingredient in animal feeds because of its high digestibility, high protein content, and unique amino acid profile. The production of soybeans—primarily in Brazil—is associated with deforestation and other environmental impacts, such as pesticide leaching. Fishmeal and fish oil are the most nutritious and digestible ingredients for farmed fish feeds. A significant proportion of world fisheries production is processed into fishmeal and fish oil. However, nearly 90 percent of the world's marine fish stocks are now depleted, overexploited, or exhausted (UNCTAD 2018). In 2014, more fish were farm raised than wild caught, and over the next decade, 67 percent of fish will be farm raised. That said, breakthroughs in fish breeding and the adoption of high-quality pelleted feeds have made large-scale, farm-raised aquaculture production possible.

Large-scale insect farming is an environmentally friendly way to take the pressure off soy and fishmeal feed protein. The environmental unsustainability of soy and fishmeal production is driving the search for alternative protein sources for animal feed. Insects are increasingly recognized as a viable alternative. Using insects in animal feeds has been researched for decades (Makkar et al. 2014). This is because insects are a natural component of many bird and mammal diets, including free-ranging chickens; pigs, which by nature are omnivorous; and most cultured fish species, such as carp, salmon, and tilapia. BSFL and mealworms have nutrient contents that are comparable to traditional feed ingredients and have the potential to replace up to 100 percent of the fishmeal feed used in pork, poultry, and aquaculture production. Farmers have several motivations for wanting to replace fishmeal with other proteins. These motivations include fishmeal's high prices, poor quality, and low supply in local markets (Pomalégni et al. 2018). Or, as one male farmer in the Arab Republic of Egypt remarked, "By replacing fishmeal and soybeans with insect meal, we are solving critical problems at the very start of the food chain. This means that ingredients used for animal and aquaculture feed do not compete with food for human consumption." In general, replacing fish and soybean meal with insect meal in fish and chicken diets does not negatively affect the fish or chicken's flavor, juiciness, or texture (Ssepuuya et al. 2019).

Waste Conversion

Insects' ability to convert feed substrate into edible body mass leads to environmental benefits. Some insects, most notably BSFL, are capable of consuming most organic waste, including that of humans and animals. This makes the BSF an ideal candidate for breeding in African FCV countries where organic waste management is a challenge.[11] In such cases, it is still important to avoid waste substrates with high levels of heavy metals. Using BSFL for feed and composting can reduce GHG emissions. A study in Indonesia found that composting segregated kitchen waste with BSFL can reduce direct CO_2-eq emissions by 47 times (Mertenat, Diener, and Zurbrügg 2019). It also found that organic waste composting with BSFL as opposed to open-air composting reduces GWP by half.

Farmed insect species convert the organic substrate they feed on very efficiently compared with conventional livestock. The growth efficiency of farmed animals is expressed as the FCR. The FCR evaluates how much feed substrate is needed to produce 1 kg of meat. Insect species have the potential of efficient growth with an FCR as low as 1.4. This is well below the FCR for chicken, which has the most optimized FCR among traditional livestock species (table 3.9). Insects can efficiently convert low-grade organic waste into high-quality fat and protein. When insects are dried, up to 70 percent of their dry matter is protein

TABLE 3.9 Feed Conversion Rates of Various Insect and Livestock Species

Species	Feed conversion ratio	Description of the farming system	Reference
Cricket (*Acheta domestica* and *Gryllus bimaculatus*)	1.82	Sheltered, open-walled system (Thailand)	Halloran et al. 2017
House cricket (*Acheta domesticus*)	2.3–6.1	Experiments in a laboratory setting	Oonincx et al. 2015
Black soldier fly (*Hermetia illucens*)	1.4–2.6	Experiments in a laboratory setting	
Mealworm (*Tenebrio molitor*)	3.8–19.1	Experiments in a laboratory setting	
Swine	4.04	National average (United States)	Mekonnen et al. 2019
Broiler chicken	2.68	National average (United States)	
Layer chicken	2.26	National average (United States)	
Turkey	3.58	National average (United States)	
Beef cow	23.5	National average (United States)	
Striped catfish (*Pangasianodon hypophthalmus*)	1.57	Intensive farming systems (Vietnam)	Hasan and Soto 2017

Source: Original table for this publication.

(Rumpold and Schluter 2013) and, similar to other foods of animal origin, the protein is rich in essential amino acids (Rumpold and Schluter 2013; Makkar et al. 2014). Therefore, insects contribute high protein quality and important micronutrients to human diets (Rumpold and Schluter 2013) and animal feeds (Makkar et al. 2014).

Insects' low FCR is partly why insect farming has fewer climate and environmental impacts than traditional livestock farming. A low FCR indicates higher efficiency in producing a kilogram of meat, or milk in the case of dairy cows. The FCR is calculated in different ways for different animals, so caution should be taken when directly comparing different animals' FCRs. Table 3.9 shows the FCR values for different insect production systems. The table indicates promising efficiencies from insect farming despite insect farming systems being in their infancy and not yet fully optimized. For example, livestock species with a long history of animal husbandry have significantly decreased their FCRs over time; between 1944 and 2017, the FCR of dairy cows for milk production dropped by 77 percent (Mekonnen et al. 2019).

BSFL have the biological capacity for efficient conversion of a broad range of organic substrates to food and protein (Li et al. 2011; St-Hilaire et al. 2007; Tinder et al. 2017; Chen et al. 2019). These substrates include vegetable waste, mixed household waste, animal waste such as manure or slaughter offal, and industrial waste from breweries, wineries, or other industries (table 3.10). These wastes and by-products increase larval performance and nutritional composition (Meneguz et al. 2018). The type of feed substrate can also affect the BSFL's growth rate and nutritional composition. Higher protein in the feed substrate generally translates into greater growth and protein content of the larvae. For example, brewery waste contributes low-grade protein to the feed substrate, which converts into high-quality protein in the BSFL. Studies show that even small amounts of brewery waste were effective at increasing larval performance (Meneguz et al. 2018). A Ghanaian farmer said, "The hatched larvae of the black soldier fly help to break down domestic and industrial waste and contribute to keeping our communities clean and safe for all."

Insects can reduce organic waste in urban settings by converting it into high-quality protein (Nyakeri et al. 2019; Nyakeri et al. 2016, 2017). This is especially valuable in high-density urban areas with poor sanitation and human waste management. In these areas, insects address health problems by reducing the amount of human waste (Banks, Gibson, and Cameron 2014). In Africa, two companies, BioCycle and Sanergy, use BSFL to break down human waste and create animal feed. The larvae that feed on this waste are always heat treated to kill any pathogens prior to using them for animal feed.

Insects convert all sorts of waste substrates into protein. Three-quarters of farms in the farm-level survey used mixed organic waste substrates. Nearly half (43 percent) of the farmers using organic waste substrates used household waste, such as vegetable peels and similar waste. Table 3.10 shows the types of substrates farms use to feed domesticated insects. The high diversity of substrates demonstrates the potential of insects to convert all sorts of low-value organic

TABLE 3.10 Substrate Use on African Insect Farms

Substrate category (*n* = number of farmers responding)	Total number of farms utilizing this substrate (% of respondents)	Origin of the substrate	Farmers using the substrate category on their farms (%)
Mixed organic waste (*n* = 154)	115 (74.7)	Household	43.0
		Catering	28.5
		Other	28.5
Food waste from shops and markets (*n* = 151)	122 (80.8)	Plant origin	79.0
		Animal origin	21.0
Industrial food waste (*n* = 153)	137 (89.5)	Brewery	64.0
		Miller	18.0
		Bakery	9.0
		Other	9.0
Commercial animal feed (*n* = 158)	48 (30.4)	Chicken feed	98.0
		Other	2.0
Manure (*n* = 154)	18 (11.7)	Animal origin	100.0
		Human origin	0
Other (*n* = 161)	66 (40.0)	Miscellaneous sources	—

Source: Original table for this publication, using the report's farm-level survey in 2019.
Note: Human waste substrates were not covered by the farm survey. — = not applicable.

waste streams into protein. Most of the farms also used industrial food waste, mainly brewery waste (64 percent of the farms using industrial food waste used brewery waste). Twelve percent of insect farmers used animal manure waste as substrate. Finally, 30 percent of farms, mainly cricket farms, used commercial compound animal feed, such as chicken feed, as substrate. Other substrates include mulberry leaves for silkworm production, palm tree trunks and palm fiber for palm weevil larvae production, and leaves from fresh plants—such as yam, pumpkin, cassava, and moringa trees—for other insects.

Natural Biofertilizer

Insect biofertilizer can improve soil health. Many parts of Africa have arid or low-fertility soils. Frass, or insect manure, can be used as an organic biofertilizer for crop soils, replacing harmful chemical fertilizers. These insect biofertilizers are rich in carbon, which can revitalize low-quality soils, or as one Congolese farmer said, "The mopane caterpillar's manure fertilizes soil. The plants that use it are regenerated." The few studies that examine replacing chemical fertilizers with insect biofertilizers show promise in the practice. An older farmer from Madagascar voiced this promise by saying, "We obtain a very good fertilizer [from silkworm frass], more efficient than chemical fertilizer." A study on Thai cricket farms found that cricket biofertilizer has higher percentages of nitrogen, phosphorus, and potassium than broiler chicken manure fertilizer (table 3.11) (Halloran et al. 2017). A Canadian study found that BSFL biofertilizers increased crop yields (Temple et al. 2013). Informal interviews for this report with insect farmers in Asia and Africa found anecdotal evidence that insect biofertilizer can improve crop growth. Biofertilizer is becoming a commercial product, especially as BSF production is growing rapidly.

Endangered Species Protection

Wild harvested insect populations are threatened. Over 40 percent of all insect species—not necessarily edible species—are currently threatened with extinction (Sánchez-Bayo and Wyckhuys 2019). Land use changes—including agriculture-related habitat loss and forest fragmentation—and the inappropriate use of agrochemicals threaten insect species (Yen 2009). The increased commercialization of insects, driven by higher demand and higher prices, could also potentially overexploit wild harvested insects if they are not adequately managed (Illgner and Nel 2000). This overharvesting could also reduce insects' larval production, compromising future generations (Langley et al. 2020). The volume of insects harvested from natural habitats fluctuates by season. This is exemplified by the shea caterpillar in Burkina Faso (Bama et al. 2018), the mopane worm in Zimbabwe (Hope et al. 2009), and several wild harvested

TABLE 3.11 Nitrogen, Phosphorus, and Potassium Content of Chicken, Cricket, and Black Soldier Fly Larvae Biofertilizers

Element	Farmed species	Nitrogen, phosphorus, and potassium in biofertilizer (%)
Nitrogen	Field cricket (*Gryllus bimaculatus*)	2.58
	House cricket (*Acheta domesticus*)	2.27
	Black soldier fly (*Hermetia illucens*)	4.66
	Broiler chicken (CP Brown)	1.70
Phosphorous	Field cricket (*Gryllus bimaculatus*)	1.55
	House cricket (*Acheta domesticus*)	2.02
	Black soldier fly (*Hermetia illucens*)	2.40
	Broiler chicken (CP Brown)	1.33
Potassium	Field cricket (*Gryllus bimaculatus*)	1.78
	House cricket (*Acheta domesticus*)	2.26
	Black soldier fly (*Hermetia illucens*)	3.00
	Broiler chicken (CP Brown)	1.58

Sources: Halloran 2017; Temple et al. 2013.

species in southeastern Nigeria (Ebenebe et al. 2017). Drought and climate change further stress insect habitats (Krug 2017).

Insect farming can alleviate pressure on overhunted, endangered species. Many traditional groups in Africa still hunt wild "bushmeat." Bushmeat is an important food source in war-torn areas of Africa with poor infrastructure where agricultural commodities cannot reach rural communities (Cawthorn and Hoffman 2015). In some Malagasy villages, 75 percent of ASF is derived from forest-dwelling animals, like lemurs. Approximately 94 percent of lemur species are threatened with extinction. Child malnutrition tends to be higher in households that hunt lemurs, indicating that these families may turn to hunting bushmeat in the absence of lemurs (Borgerson et al. 2017). That said, conservationists cannot simply ban bushmeat hunting because if no nutritious substitutes are accessible, the number of children suffering from anemia could increase by 29 percent and anemia cases could triple among children in the poorest households (Golden et al. 2011). Insect farming could alleviate some of the demand for bushmeat (Yen 2015). In Madagascar, insect farming is being promoted to reduce the pressure on wildlife species that are hunted when other foods are less abundant. Likewise, insect farming can also alleviate pressures on wild habitats. For example, palm weevils feed on raffia stems in natural forests in Africa. In Cameroon, African palm weevil farming meant that fewer raffia stems were exploited during traditional weevil collection practices. As a result, weevil farming helped assure the survival and sustainability of the area's raffia forest ecosystem (Muafor et al. 2015). Insects alone will not be a panacea for the bushmeat crisis. Yet they could play a very significant role in alleviating the demand for vertebrate bushmeat if regionally specific solutions are considered.

Proper insect management can limit degradation caused by wild insect harvesting. According to the country-level survey, a forest management project in Madagascar aims to conserve 203 hectares of tapia (*Uapaca bojeri*) forest and the surrounding flora and fauna. These natural areas are being degraded by the wild harvest of silkworms. The project encloses part of the tapia trees with metal mesh fencing to deter mismanagement. Wild silkworm larvae (*Borocera cajani*) are then placed directly on the tapia trees in the enclosure. The local communities harvest the silkworms from the enclosures. After three years, the enclosure is removed and the silkworms are released to repopulate and sustain the wild population. This project protects the tapia forest and produces wild silkworm chrysalises to support the livelihoods of 34 households in the local community.

Reducing the reliance on wild bushmeat can also directly benefit human health (Bett et al. 2018). Recent epidemics and pandemics, such as Ebola and COVID-19 (coronavirus), demonstrate how wild animals can be reservoirs for zoonotic diseases, which jump from animals to humans. Approximately 70 percent of emerging infectious diseases in humans over the past 30 years are of zoonotic origins (Wang and Crameri 2014). Replacing wild bushmeat hunting with insect farming can reduce the risk of such diseases.

INSECT FARMING'S ECONOMIC BENEFITS

This section looks at the prices and profitability of the insect farming sector. Forty-four percent of the surveyed farmers said income generation was why they started raising insects. A Malagasy farmer said, "[Silkworm farming] is important for us because it allows us to cover our needs during the lean season." This section finds that the insect industry is growing because of its profitability, especially for small-scale farmers. Prices remain high, but they will decline as the industry matures. As a result, insect farming is a new industry with promising economic potential in Africa.

Profitability

Insect farming can be a very profitable agricultural activity for small-scale farmers. In Cameroon, the average income of formal African palm weevil larvae collectors varied between US$180 and US$600 per month, which was 30 to 75 percent of their household income (Muafor et al. 2015). Palm weevil incomes were higher than bushmeat hunting incomes. Incomes from palm weevil sales were also significantly higher than the incomes of unskilled workers or rural coffee producers (Muafor et al. 2015). In Ghana, economic viability analyses show that African palm weevil farmers could pay back their initial capital investment in weevil farming in 127 days. This requires selling 3,020 weevil larvae at US$0.06 per larva. In a year, a farmer could have three production cycles and generate total revenue of US$553, which would require selling 755 larvae per month. This would produce a net cash availability of US$265.25 and a net profit of US$82.16 in the first year of production. In Côte d'Ivoire, vendors of wild, unfarmed insects have an estimated average monthly income of US$98.50 (Ehounou, Ouali-N'goran, and Niassy 2018). In Thailand, small-scale cricket farming has diversified rural household incomes, and the average incomes can be significantly more than those from other agricultural activities, especially in the northeastern part of the country (Halloran, Roos, and Hanboonsong 2017). Contract farmers working for one of Thailand's largest cricket companies earn profits of US$1,950 per year, which is higher than other agricultural work. Independent cricket farmers earn similar profits (table 3.12). In Korea, insect farming revenues vary greatly among insect species but are much higher than minimum wage earnings, which were US$7 per hour in 2020 (table 3.13). The research team's informal conversations with rural residents in Thailand and Korea suggest that insect farming can be an attractive livelihood for urban dwellers wishing to return to the countryside.

There are other profitable livelihoods associated with insect farming, for example, consultants who train and provide technical assistance to farmers or insect production companies. A Ghanaian farmer said, "Insect farming serves as a secondary income generation activity for me. Consulting on insect farming alone gives me almost half of my annual salary as a government employee." Tourism associated with insect farming has taken place in some areas as well. In Thailand, for example, tourist maps guide visitors to cricket farming areas and

TABLE 3.12 Annual Production for a Cricket Farm of Eight Pens with 8.5 Annual Growth Cycles in Thailand

2019 US dollars

Input/ output	Unit	Amount per unit	Price per unit (US$)	Units per cycle per pen	Expenses per year (US$)	Revenue (US$)	Profit (US$)
Feed	kg	30	15.04	2	2,045		
Recycled egg cartons	#	750	0.05	187.5	638		
Crickets	kg	1	0.44	150		4,488	
Biofertilizer	kg	30	1.28	2.5		218	
Total					2,683	4,706	2,023

Source: Original table for this publication, using data from 2019.
Note: # = number; kg = kilograms.

TABLE 3.13 Annual Revenue from Insect Farming in the Republic of Korea

2020 US dollars

Insect	Number of laborers required	Output (kg)	Annual revenue (US$)	Annual production costs (US$)	Annual profit (US$)
Mealworms	2	3,500	167,000	67,000	100,000
White-spotted flower chafer	1–2	3,000	33,000	17,000	16,000
Rhinoceros beetle	1–2	150	41,700	25,000	16,700
Crickets	1–2	3,000	37,500	17,000	20,500
Black soldier fly	1	24,000	320,000	6,000	314,000

Source: RDA 2020b.
Note: kg = kilograms.

tour operators provide guided visits to insect farms (Halloran, Roos, Flore, and Hanboonsong 2016). And insect farming affords time for other profitable activities. For example, in Thailand, small- and medium-scale cricket farmers spend only an average of 2.9 hours per day tending to their crickets, leaving them time for other profitable activities (Halloran, Roos, and Hanboonsong 2017).

Falling Prices

Insect prices vary significantly in Africa. This may reflect the insect sector's evolving status and uneven supply in most African countries. An insect farming expert in the Democratic Republic of Congo said, "The prices are not stable;

they vary from one place to another. In the vicinity of the main avenues the prices are higher than in more distant places." In most countries, insect prices are not competitive with those of other sources of animal feed protein. That said, there are some exceptions. In Kenya, BSFL prices have fallen to lower than fishmeal prices, a promising development for the insect sector. According to the farm-level survey, most farmers (55 percent) also sell insect by-products, like biofertilizer and insect compost, and half (49 percent) of the farmers sell co-products, like insect eggs, nymphs, and larvae. All that being said, farmers often do not sell their insects at all, instead consuming them themselves or feeding them to on-farm livestock. Or as a farmer in the Democratic Republic of Congo said, "It's easy [to farm insects] and you can do it all year. If it does not sell, you can eat [it] yourself." According to the farm-level survey, 48 percent of farmers sell all the insects they produce, 25 percent sell half of what they produce, and 27 percent sell less than a quarter of what they produce. Table 3.14 shows the prices for different types of insects, and table 3.15 shows the prices for the various insect by-products and co-products.

The prices for farmed insects are high and climbing, but they are likely to decline as the industry becomes more established. In Africa, farmed insect prices are still relatively high. For example, in Ghana, fresh African palm weevil larvae cost US$10/kg, compared with US$3/kg for chicken, US$2.80/kg for fish, and US$2.40/kg for boned beef. In Angola, the price of wild palm weevil larvae is approximately 9.5 times higher than that of dry beans (Lautenschläger et al. 2017). These prices are expected to rise because there are too few producers to meet the regional demand.[12] For example, in Cameroon in the late 1990s, US$0.20 purchased 12 wild African palm weevil larvae but now purchases only four (Muafor et al. 2015). In Yaoundé, 43 wild larvae sell for US$3.00, while three or four roasted larvae sell for US$0.20—all relatively high prices. Insect prices will likely drop as the edible insect market matures. In Finland, prices of farmed cricket powder halved from US$110/kg in 2017 to US$55/kg in 2019. In Thailand, the price for the same product fell from US$33/kg to US$22/kg across the same period (Barclays Bank 2019).

Farmers say the rise in prices is caused by increased consumer demand for insects. According to the farm-level survey, 59 percent of the surveyed farmers said insect selling prices had risen from the previous year. Of these farmers, 23 percent said the price rose less than 10 percent, 47 percent said the price rose by 11 to 25 percent, 16 percent said the price rose by 25 to 50 percent, and 14 percent said the price rose by more than 50 percent. The farmers cited greater demand for insects coupled with short supplies as contributing to this price increase. An Egyptian farmer explained this by saying, "We see that as the awareness for the product increases, the demand for the product increases too. More interested customers are reaching out." For example, 66 percent of farmers said the demand for insects has increased from the previous year. Thirty-three percent of farmers said demand increased by 10 percent, 32 percent said it increased by 11 to 25 percent, 9 percent said it increased by 26 to 50 percent,

TABLE 3.14 Estimated Prices and Volumes of Selected Farmed Insect Species

Common name (units)	Scientific name	Price range (US$/unit)			Average price ($US/unit)			Estimated total annual volume produced (tons of fresh insects)
		Fresh	Dried	Powdered	Fresh	Dried	Powdered	
Black soldier fly larvae (tons)	*Hermetia illucens*	940.5–1,800	600–2,639	2,990	1,168	1,913	2,990	293
Housefly larvae (tons)	*Musca domestica*	2,000–7,000	n.a.	n.a.	4,500	n.a.	n.a.	15
Field crickets (kg)	*Gryllus bimaculatus*	0.50–7,802	10.00–29.70	10.00–30.00	1,308	23.10	17.43	2.3–8.5
House crickets (kg)	*Acheta domesticus*	9.90–13.43	10.00–15.00	10.00–30.00	12.07	12.50	20.00	1.1–2.5
Silkworm chrysalis (kg)	*Bombyx mori*	0.86–6.89	n.a.	n.a.	3.62	n.a.	n.a.	24.1
African palm weevil larvae (kg)	*Rhynchophorus phoenicis*	3.20–31.73	n.a.	n.a.	7.80	n.a.	n.a.	2.2
Mopane caterpillars (kg)	*Gonimbrasia belina*	1.50–3.00	n.a.	n.a.	2.25	n.a.	n.a.	1.0

Source: Original table for this publication, using the report's farm-level survey in 2019.
Note: kg = kilograms; n.a. = not applicable.

TABLE 3.15 Examples of By-products and Co-products Sold by African Insect Farms

Common name	Scientific name	By-product or co-product	Price range (US$/unit)	Average price (US$/unit)
Black soldier fly larvae	*Hermetia illucens*	Biofertilizer	150/ton	150/ton
		Eggs (approximately 500 grams on a plate)	5.94–24.75/500 grams	15.00/ 500 grams
Field crickets	*Gryllus bimaculatus*	Eggs (approximately 3,000 eggs on a plate)	2.66–4.95/ plate	3.81/plate
		Parent stock (males and females for starting a new farm)	4.00–5.44/ container (120 females and 50 males)	5.08/ container
House crickets	*Acheta domesticus*	Eggs (approximately 3,000 eggs on a plate)	1.33–4.95/ plate	2.60/plate
		Parent stock (males and females for starting a new farm)	5.32–6.66/ container (120 females and 50 males)	5.77/ container

Source: Original table for this publication, using on the report's farm-level survey in 2019.

and 26 percent said it increased by over 50 percent. Farmers attributed this demand to the needs of consumers (45 percent), middlepersons (42 percent), and other sources (13 percent).

Insect feed prices are becoming more competitive with soy and fishmeal. Currently, the production volumes of fishmeal, soybean extract, and soybean meal are far higher than that of insect meal, allowing for cheaper prices. Informal interviews carried out for this report approximate that BSFL whole meal (not defatted) sells for US$1/kg in Asia, US$1.80/kg in Europe, and US$2/kg in North America. A study in Kenya found that using BSFL meal in pig diets did not affect the profitability of the pigs (Chia et al. 2019). The cost-benefit ratio and return on investment did not differ among diets that replaced fishmeal with BSFL meal. This shows that BSFL meal is a beneficial substitution as a component of pig feed from a performance-based perspective and an economic-based perspective. Another project in Kenya and Uganda found that using insect feed could reduce the cost of protein from livestock rearing and aquaculture by 25.0 to 37.5 percent in the short term and by 41.7 to

51.4 percent in the medium term (Fiaboe and Nakimbugwe 2017). Another study from West Africa showed that BSFL feed produced from low-value substrates could achieve prices comparable to fishmeal (Roffeis et al. 2020). This is because the economic value of the waste used as BSFL substrate is 100 to 200 times lower than the value of the larvae.

Other evidence suggests that insect feeds are still not as economically viable as other feeds but could soon be in the future. A study on insect-based feeds in West Africa shows that economic performance is largely determined by the costs of labor and procuring rearing substrates (Roffeis et al. 2018). The study found that housefly larvae production is more economically advantageous than BSFL production. The breakeven price is US$1.41–US$1.91/kg for dry insect-based feed from housefly larvae and US$2.79/kg for dry insect-based feed from BSFL. The study showed that insect-based feeds can replace imported fishmeal but do not have an advantage over plant-based feeds like soybean meal. Another study on sea bass found that introducing insect meal would increase feeding costs because of the high market prices for mealworm flour and a less competitive FCR than that of fishmeal (Arru et al. 2019). However, it is expected that insect feed prices will become more competitive as the industry matures and reaches scale.

ANNEX 3A

TABLE 3A.1 Insect Consumption in African Fragility-, Conflict-, and Violence-Affected States

Insect family	Burkina Faso	Burundi	Cameroon	Central African Republic	Chad	Conmoros	Congo, Dem. Rep.	Congo, Rep.	Eritrea	Gambia, The	Guinea-Bissau	Liberia	Libya	Mali	Mozambique	Niger	Nigeria	Somalia	South Sudan	Sudan	Zimbabwe
Acrididae	2		29	6			2	16					1	4		19	1		1	4	8
Anostostomatidae								1													2
Aphididae																					
Apidae				11			9										1				
Belostomatidae								1													
Blattidae			1																		
Bombycidae																					
Bostrichidae								1													
Brahmaeidae			1				1	1													
Buprestidae			1																		1
Cerambycidae				3			4								1		1				
Ceratocampidae							1														
Cercopidae							6														

(Continued)

TABLE 3A.1 Insect Consumption in African Fragility-, Conflict-, and Violence-Affected States *(Continued)*

Insect family	Burkina Faso	Burundi	Cameroon	Central African Republic	Chad	Conmoros	Congo, Dem. Rep.	Congo, Rep.	Eritrea	Gambia, The	Guinea-Bissau	Liberia	Libya	Mali	Mozambique	Niger	Nigeria	Somalia	South Sudan	Sudan	Zimbabwe
Cicadidae							2	6													1
Coreidae																					3
Crambidae																	1				
Curculionidae																					
Dinidoridae																				1	1
Dryophthoridae			1	1			1	1									1				
Dytiscidae							1														
Elateridae																					
Formicidae			3		1		1	1							1					1	2
Gryllacridae																					
Gryllidae				2			2	1			1	1					2				1
Gryllotalpidae																					1
Hesperiidae								1													
Hydrophilidae																					
Lasiocampidae				1																	1

(Continued)

TABLE 3A.1 Insect Consumption in African Fragility-, Conflict-, and Violence-Affected States *(Continued)*

Insect family	Burkina Faso	Burundi	Cameroon	Central African Republic	Chad	Conmoros	Congo, Dem. Rep.	Congo, Rep.	Eritrea	Gambia, The	Guinea-Bissau	Liberia	Libya	Mali	Mozambique	Niger	Nigeria	Somalia	South Sudan	Sudan	Zimbabwe
Libellulidae																					
Limacodidae							1														1
Lucanidae																					
Lymantriidae							1														
Mantidae																					
Nephilidae										1											
Nepidae							1														
Noctuidae							2	1													
Notodontidae T.				1			3	1									3				1
Notodontidae				2			9														
Nymphalidae				2			2														
Odonata			1														1				
Palingeniidae																					
Papilionidae				1																	
Passalidae																					

(Continued)

TABLE 3A.1 Insect Consumption in African Fragility-, Conflict-, and Violence-Affected States *(Continued)*

Insect family	Burkina Faso	Burundi	Cameroon	Central African Republic	Chad	Conmoros	Congo, Dem. Rep.	Congo, Rep.	Eritrea	Gambia, The	Guinea-Bissau	Liberia	Libya	Mali	Mozambique	Niger	Nigeria	Somalia	South Sudan	Sudan	Zimbabwe
Pentatomidae																	1			1	
Phasmatidae																					
Psychidae			1	1			2	2													
Pyrgomorphidae			3	2			1	3							1		1				1
S.Cetoniinae			2	1			2														
S.Dynastinae			2	1			2	3									2				
S.Melolonthinae			2																		3
S. Rutelinae			2																		
S.Trichiinae				1			1														
Saturniidae	1		3	7			54	12						2			3				10
Scaradaeidae																					
Sphecidae							1														
Sphingidae				4			4	1													
Tarachodidae			1																		
Tenebrionidae								1													

(Continued)

TABLE 3A.1 Insect Consumption in African Fragility-, Conflict-, and Violence-Affected States *(Continued)*

Insect family	Burkina Faso	Burundi	Cameroon	Central African Republic	Chad	Conmoros	Congo, Dem. Rep.	Congo, Rep.	Eritrea	Gambia, The	Guinea-Bissau	Liberia	Libya	Mali	Mozambique	Niger	Nigeria	Somalia	South Sudan	Sudan	Zimbabwe
Termitidae	1	1	3	5			8	2			1	1		1			2				4
Tessaratomidae																					3
Tettigoniidae			1	2			2	1													
Thespidae			1																		
Vesipdae							1														
Total count	***4***	***1***	***58***	***54***	***1***	***0***	***127***	***57***	***0***	***1***	***2***	***2***	***1***	***7***	***3***	***19***	***20***	***0***	***1***	***7***	***44***

Source: Adapted from Jongema 2017.

Note: The numbers in the table represent the number of insect species in each indicated insect family consumed in the designated country.

NOTES

1. The focus of this report is on African FCV countries, but the issues discussed in this section are applicable to FCV-affected states outside Africa in which insects are commonly consumed.
2. This includes insects produced as pets.
3. https://protix.eu/.
4. https://flyfarm.com/.
5. EFSA opinion on locusts (https://efsa.onlinelibrary.wiley.com/doi/abs/10.2903/j.efsa.2021.6667) and opinion on mealworms (https://www.efsa.europa.eu/en/efsajournal/pub/6343).
6. https://eur-lex.europa.eu/legal-content/EN/TXT/?uri=CELEX%3A32021R0882&qid=1622617276506.
7. Regulation (EC) No 999/2001: Prohibition to feed non-ruminant farmed animals, other than fur animals, with protein derived from animals (https://eur-lex.europa.eu/legal-content/EN/TXT/PDF/?uri=OJ:L:2021:295:FULL&from=EN).
8. Regulation (EC) No 999/2001: Prohibition to feed non-ruminant farmed animals, other than fur animals, with protein derived from animals (https://eur-lex.europa.eu/legal-content/EN/TXT/PDF/?uri=OJ:L:2021:295:FULL&from=EN).
9. Compound animal feed is blended from various raw materials.
10. The guidelines are available as an e-book at http://lib.rda.go.kr.
11. The unique characteristics of BSF in the food chain are presented in the next section, on insect farming's economic benefits.
12. J. Anankware, personal communication, 2019.

REFERENCES

360 Market Updates. 2019. "Global Insect Feed Market 2019 by Manufacturers, Regions, Type and Application." Forecast 2024 (accessed February 2020). https://www.360marketupdates.com/global-insect-feed-market-13620651.

ACFS (National Bureau of Agricultural Commodity and Food Standards). 2019. "Good Agricultural Practices for Cricket Farming." ACFS, Ministry of Agriculture and Cooperatives, Bangkok, Thailand.

AgriProtein. 2018. "USD 105 Million Raised for Sustainable Feed Firm: AgriProtein Secures Largest Investment to Date in Insect Protein Sector." Press Release, AgriProtein Online, June 4, 2018. https://www.agriprotein.com/press-articles/usd-105-million-raise-for-sustainable-feed-firm/.

AgriTech Capital. 2019. "The Buzz about Insect Protein." AgriTech Capital, June 22, 2019. https://www.agritechcapital.com/ideas-you-can-use-food-beverage/2019/6/22/the-buzz-about-insect-protein.

Alemu, M. H., and S. B. Olsen. 2018. "Kenyan Consumers' Experience of Using Edible Insects as Food and Their Preferences for Selected Insect-Based Food Products." In *Edible Insects in Sustainable Food Systems*, edited by A. Halloran, R. Flore, P. Vantomme, and N. Roos, 363–74. Springer International Publishing. https://doi.org/10.1007/978-3-319-74011-9_22.

Aniebo, A. O., E. S. Erondu, and O. J. Owen. 2009. "Replacement of Fish Meal with Maggot Meal in African Catfish (Clarias gariepinus) Diets." *Revista Científica UDO Agrícola* 9: 666–71.

Arru, B., R. Furesi, L. Gasco, F. Madau, and P. Pulina. 2019. "The Introduction of Insect Meal into Fish Diet: The First Economic Analysis on European Sea Bass Farming." *Sustainability* 11: 1697.

Awoniyi, T. A. M., V. A. Aletor, and J. M. Aina. 2003. "Performance of Broiler-Chickens Fed on Maggot Meal in Place of Fishmeal." *International Journal of Poultry Science* 2: 271–74.

Ayieko, M., and V. Oriaro. 2008. "Consumption, Indigeneous Knowledge and Cultural Values of the Lakefly Species within the Lake Victoria Region." *African Journal of Environmental Science and Technology* 2: 282–86.

Bama, H. B., R. B. Dabire, D. Quattara, S. Niassy, M. N. Ba, and D. Dakouo. 2018. "Diapause Disruption in Cirina Butyrospermi Vuillet (Lepidoptera, attacidae), the Shea Caterpillar, in Burkina Faso." *Journal of Insects as Food and Feed* 4: 239–45.

Banks, I. J., W. T. Gibson, and M. M. Cameron. 2014. "Growth Rates of Black Soldier Fly Larvae Fed on Fresh Human Faeces and Their Implication for Improving Sanitation." *Tropical Medicine & International Health* 19: 14–22.

Barclays Bank. 2019. "Barclays Research Highlights: Sustainable & Thematic Investing: Food Revolution." Barclays Bank, London. https://www.investmentbank.barclays.com/content/dam/barclaysmicrosites/ibpublic/documents/our-insights/FoodWaste/Leaflet Food Revolution.pdf.

Beniers, J. J. A., and R. I. Graham. 2019. "Effect of Protein and Carbohydrate Feed Concentrations on the Growth and Composition of Black Soldier Fly (Hermetia illucens) Larvae." *Journal of Insects as Food and Feed* 5: 193–99.

Bett, B., N. Ngwili, D. Nthiwa, and A. Silvia. 2018. "Association between Land Use Change and Exposure to Zoonotic Pathogens: Evidence from Selected Case Studies in Africa." In *Encyclopedia of Food Security and Sustainability*, edited by P. Ferranti, E. M. Berry, and J. R. Anderson, 463–68. Elsevier. doi:10.1016/B978-0-08-100596-5.21573-4.

Bhutta, Z. A., T. Ahmed, R. E. Black, S. Cousens, K. Dewey, E. Giugliani, et al. 2008. "Maternal and Child Undernutrition 3: What Works? Interventions for Maternal and Child Undernutrition and Survival." *Lancet* 371: 417–40.

Black, R. E., C. G. Victora, S. P. Walker, Z. A. Bhutta, P. Christian, O. M. De, et al. 2013. "Maternal and Child Undernutrition and Overweight in Low-Income and Middle-Income Countries." *Lancet* 382: 427–51.

Borgerson, C., D. Rajaona, B. Razafindrapaoly, B. J. R. Rasolofoniaina, C. Kremen, and C. D. Golden. 2017. "Links between Food Insecurity and the Unsustainable Hunting of Wildlife in a UNESCO World Heritage Site in Madagascar." *Lancet* 389: S3.

Buhler Group. 2019. "Bühler Insect Technology Solutions and Alfa Laval Join Forces in Insect Processing." Buhler Group, Uzwil, Switzerland. https://www.buhlergroup.com/content/buhlergroup/global/en/media/media-releases/buehler_insect_technologysolutionsandalfalavaljoinforcesininsect.html.

Byrne, Jane. 2018a. "Cargill Sees an Expansion of Its Functional Fish Feed Portfolio Globally, BioMar Evaluating Novel Proteins." Feed Navigator, March 13, 2018. https://www.feednavigator.com/Article/2018/03/13/Cargill-sees-an-expansion-of-its-functional-fish-feed-portfolio-globally-BioMar-evaluating-novel-proteins.

Byrne, Jane. 2018b. "McDonald's Championing Research into Insect Feed for Chickens." Reports from Feed Protein Vision 2018, Feed Navigator, March 27, 2018. https://www.feednavigator.com/Article/2018/03/27/McDonald-s-championing-research-into-insect-feed-for-chickens.

Cawthorn, D. M., and L. C. Hoffman. 2015. "The Bushmeat and Food Security Nexus: A Global Account of the Contributions, Conundrums and Ethical Collisions." *Food Research International* 76: 906–25.

Chaalala, S., A. Leplat, and H. Makkar. 2018. "Importance of Insects for Use as Animal Feed in Low-Income Countries." In *Edible Insects in Sustainable Food Systems*, edited by A. Halloran, R. Flore, P. Vantomme, and N. Roos, 363–74. Springer International Publishing. doi:10.1007/978-3-319-74011-9_18.

Chen, J., D. Hou, W. Pang, E. E. Nowar, J. K. Tomberlin, R. Hu, et al. 2019. "Effect of Moisture Content on Greenhouse Gas and NH3 Emissions from Pig Manure Converted by Black Soldier Fly." *Science of the Total Environment* 697.

Chia, S., C. Tanga, I. Osuga, A. Alaru, D. Mwangi, M. Githinji, et al. 2019. "Effect of Dietary Replacement of Fishmeal by Insect Meal on Growth Performance, Blood Profiles and Economics of Growing Pigs in Kenya." *Animals* 9: 705.

DanChurchAid. 2020. "Where We Work: Kenya." DanChurchAid, Copenhagen, Denmark. https://www.danchurchaid.org/where-we-work/kenya.

Degrandi-Hoffman, G., H. Graham, F. Ahumada, M. Smart, and N. Ziolkowski. 2019. "The Economics of Honey Bee (Hymenoptera: Apidae) Management and Overwintering Strategies for Colonies Used to Pollinate Almonds." *Journal of Economic Entomology* 112: 2524–33.

Development Initiatives. 2018. *2018 Global Nutrition Report: Shining a Light to Spur Action on Nutrition*. Bristol, UK: Development Initiatives. https://globalnutritionreport.org/reports/global-nutrition-report-2018/.

Dror, D. K., and L. H. Allen. 2011. "The Importance of Milk and Other Animal-Source Foods for Children in Low-Income Countries." *Food and Nutrition Bulletin* 32: 227–43.

Dube, S., N. R. Dlamini, A. Mafunga, M. Mukai, and Z. Dhlamini. 2013. "A Survey on Entomophagy Prevalence in Zimbabwe." *African Journal of Food, Agriculture, Nutrition and Development* 13 (1): 7242–53.

Ebenebe, C. I., M. I. Amobi, C. Udegbala, A. N. Ufele, and B. O. Nweze. 2017. "Survey of Edible Insect Consumption in South-Eastern Nigeria." *Journal of Insects as Food and Feed* 3: 241–52.

EFSA (European Food Safety Authority). 2015. "Risk Profile Related to Production and Consumption of Insects as Food and Feed." *EFSA Journal* 13: 4257.

Ehounou, G. P., S. W. M. Ouali-N'goran, and S. Niassy. 2018. "Assessment of Entomophagy in Abidjan (Côte d'Ivoire, West Africa)." *African Journal of Food Science* 12: 6–14.

European Commission. 2019. *EU Feed Protein Balance Sheet*. Brussels, Belgium: European Commission. https://ec.europa.eu/info/sites/info/files/food-farming-fisheries/farming/documents/eu-feed-protein-balance-sheet_2017-18_en.pdf.

Evans, J., M. H. Alemu, R. Flore, M. B. Frost, A. Halloran, A. B. Jensen, et al. 2015. "'Entomophagy': An Evolving Terminology in Need of Review." *Journal of Insects as Food and Feed* 1: 293–305.

FAO (Food and Agriculture Organization of the United Nations). 2018. "Food-Based Dietary Guidelines in Africa." Food and Nutrition Division, FAO, Rome. http://www.fao.org/nutrition/education/food-dietary-guidelines/regions/africa/en/.

FAO and WHO (Food and Agriculture Organization of the United Nations and World Health Organization). 2007. *Protein and Amino Acid Requirements in Human Nutrition: Report of a Joint WHO/FAO/UNU Expert Consultation*, vol. 935. Rome: FAO.

Feng, P. F., J. Z. He, M. Lv, G. H. Huang, X. L. Chen, Q. Yang, et al. 2019. "Effect of Dietary Tenebrio Molitor Protein on Growth Performance and Immunological Parameters in Macrobrachium Rosenbergii." *Aquaculture* 511: 734247.

Fiaboe, K., and D. Nakimbugwe. 2017. *INSFEED: Integrating Insects in Poultry and Fish Feed in Kenya and Uganda*. IDRC Project Number 107839. Ottawa, Canada: International Development Research Centre.

Fischer, C. G., and T. Garnett. 2016. *Plates, Pyramids, and Planets: Developments in National Healthy and Sustainable Dietary Guidelines: A State of Play Assessment*. Rome: Food and

Agriculture Organization of the United Nations and University of Oxford Food Climate Research Network.

Fontes, T. V., K. R. B. de Oliveira, I. L. Gomes Almeida, T. M. Maria Orlando, P. B. Rodrigues, D. V. da Costa, P. V. Rosa, et al. 2019. "Digestibility of Insect Meals for Nile Tilapia Fingerlings." *Animals* 9: 181.

Food Business Africa. 2019. "AgriProtein Set for Global Expansion with New Facilities in the US and Europe." Food Business Africa, October 31, 2019. https://www.foodbusinessafrica.com/2019/10/31/agriprotein-set-for-global-expansion-with-new-facilities-in-the-us-and-europe/.

Fraval, S., J. Hammond, J. R. Bogard, M. Ng'endo, J. van Etten, M. Herrero, et al. 2019. "Food Access Deficiencies in Sub-Saharan Africa: Prevalence and Implications for Agricultural Interventions." *Frontiers in Sustainable Food Systems* 3: 104.

Ghosh, S., C. Jung, V. B. Meyer-Rochow, and A. Dekebo. 2019. "Perception of Entomophagy by Residents of Korea and Ethiopia Revealed through Structured Questionnaire." *Journal of Insects as Food and Feed* 6 (1): 59–64.

Godfray, H. C. J., P. Aveyard, T. Garnett, J. W. Hall, T. J. Key, J. Lorimer, et al. 2018. "Meat Consumption, Health, and the Environment." *Science* 361 (6399): eaam5324.

Golden, C. D., L. C. H. Fernald, J. S. Brashares, B. J. R. Rasolofoniaina, and C. Kremen. 2011. "Benefits of Wildlife Consumption to Child Nutrition in a Biodiversity Hotspot." *Proceedings of the National Academy of Sciences of the United States of America* 108: 19653–56.

Graphical Research. 2018. "APAC Edible Insects Market Growth: Industry Forecast Report 2024." Graphical Research, Maharashtra, India. https://www.graphicalresearch.com/industry-insights/1044/asia-pacific-edible-insects-market.

Hallberg, L., M. Hoppe, M. Andersson, and L. Hulthen. 2003. "The Role of Meat to Improve the Critical Iron Balance during Weaning." *Pediatrics* 111: 864–70.

Halloran, A. 2017. "The Impact of Cricket Farming on Rural Livelihoods, Nutrition and the Environment in Thailand and Kenya." PhD thesis, Department of Nutrition, Exercise and Sports, University of Copenhagen, Denmark. doi:10.13140/RG.2.2.23820.00641.

Halloran, A. 2018. "The Social Impacts of Using Black Soldier Flies for Bioconversion: A Preliminary Assessment from South Africa." GREEiNSECT report. Department of Nutrition, Exercise and Sports, University of Copenhagen, Denmark. https://greeinsect.ku.dk/publications/more-presentations/The_social_impact_of_using_black_soldier_flies_for_bioconversion.pdf.

Halloran, A., R. Flore, P. Vantomme, and N. Roos. 2018. "Introduction." In *Edible Insects in Sustainable Food Systems*, edited by A. Halloran, R. Flore, P. Vantomme, and N. Roos. Springer International Publishing. doi:10.1007/978-3-319-74011-9.

Halloran, A., Y. Hanboonsong, N. Roos, and S. Bruun. 2017. "Life Cycle Assessment of Cricket Farming in North-Eastern Thailand." *Journal of Cleaner Production* 156: 83–94.

Halloran, A., N. Roos, J. Eilenberg, A. Cerutti, and S. Bruun. 2016. "Life Cycle Assessment of Edible Insects for Food Protein: A Review." *Agronomy for Sustainable Development* 36.

Halloran, A., N. Roos, R. Flore, and Y. Hanboonsong. 2016. "The Development of the Edible Cricket Industry in Thailand." *Journal of Insects as Food and Feed* 2: 91–100.

Halloran, A., N. Roos, and Y. Hanboonsong. 2017. "Cricket Farming as a Livelihood Strategy in Thailand." *Geographical Journal* 183: 112–24.

Halloran, A., P. Vantomme, Y. Hanboonsong, and S. Ekesi. 2015. "Regulating Edible Insects: The Challenge of Addressing Food Security, Nature Conservation, and the Erosion of Traditional Food Culture." *Food Security* 7: 739–46.

Hasan, M. R., and D. Soto. 2017. *Improving Feed Conversion Ratio and Its Impact on Reducing Greenhouse Gas Emissions in Aquaculture*. Rome: Food and Agriculture Organization of the United Nations.

Heuel, M., C. Sandrock, A. Mathys, M. Gold, C. Zurbrügg, M. Kreuzer, M. Terranova, et al. 2019. "Performance of Laying Hens When Replacing Soybean Cake and Oil by Insect Larval Protein Meal and Fat." In *Energy and Protein Metabolism and Nutrition*, edited by M. L. Chizzotti, 179–80. Wageningen Academic Publishers. doi:10.3920/978-90-8686-891-9_29.

Holland, H. 2019. "From Poo to Food: Kenyan Toilet Waste Key for New Animal Feed." Reuters, June 17, 2019. https://www.reuters.com/article/us-kenya-insects/from-poo-to-food-kenyan-toilet-waste-key-for-new-animal-feed-idUSKCN1TI1BN.

Hope, R. A., P. G. H. Frost, A. Gardiner, and J. Ghazoul. 2009. "Experimental Analysis of Adoption of Domestic Mopane Worm Farming Technology in Zimbabwe." *Development Southern Africa* 26: 29–46.

Hotz, C., G. Pelto, M. Armar-Klemesu, E. F. Ferguson, P. Chege, E. Musinguzi, et al. 2015. "Constraints and Opportunities for Implementing Nutrition-Specific, Agricultural and Market-Based Approaches to Improve Nutrient Intake Adequacy among Infants and Young Children in Two Regions of Rural Kenya." *Maternity and Child Nutrition* 11: 39–54.

IAEA (International Atomic Energy Agency). 2020. "Sterile Insect Technique." IAEA, Vienna, Austria. https://www.iaea.org/topics/sterile-insect-technique.

IFIF (International Feed Industry Federation). 2019. "Global Feed Statistics." IFIF, Wiehl, Germany. https://ifif.org/global-feed/statistics/.

IFPRI (International Food Policy Research Institute). 2015. *2015 Global Nutrition Report: Actions and Accountability to Advance Nutrition & Sustainable Development*. Washington, DC: IFPRI. doi:10.2499/9780896298835.

Illgner, P., and E. Nel. 2000. "The Geography of Edible Insects in Sub-Saharan Africa: A Study of the Mopane Caterpillar." *Geographical Journal* 166: 336–51.

IPIFF (International Platform of Insects for Food and Feed). 2020. "EU Legislation." IPIFF, Brussels, Belgium. http://ipiff.org/insects-eu-legislation/.

IPIFF (International Platform of Insects for Food and Feed). 2021. "An Overview of the European Market of Insects as Feed." IPIFF, Brussels, Belgium. https://ipiff.org/wp-content/uploads/2021/04/Apr-27-2021-IPIFF_The-European-market-of-insects-as-feed.pdf.

Irungu, F. G., C. M. Mutungi, A. K. Faraj, H. Affognon, S. Ekesi, D. Nakimbugwe, and K. K. M. Fiaboe. 2018. "Proximate Composition and *In Vitro* Protein Digestibility of Extruded Aquafeeds Containing *Acheta Domesticus* and *Hermetia Illucens* Fractions." *Journal of Insects as Food and Feed* 4: 275–84.

Jackson, L. 2020. "World's Largest Fly Factory Attracting Investors Eyeing Aquafeed Expansion." Aqua Culture Alliance, Portsmouth, NH. https://www.aquaculturealliance.org/advocate/worlds-largest-fly-factory-attracting-investors-eyeing-aquafeed-expansion/.

Jensen, L. D., R. Miklos, T. K. Dalsgaard, L. H. Heckmann, and J. V. Nørgaard. 2019. "Nutritional Evaluation of Common (Tenebrio molitor) and Lesser (Alphitobius diaperinus) Mealworms in Rats and Processing Effect on the Lesser Mealworm." *Journal of Insects as Food and Feed* 5 (4): 257–66. doi:10.3920/jiff2018.0048.

Ji, Y. J., H. N. Liu, X. F. Kong, F. Blachier, M. M. Geng, Y. Y. Liu, and Y. L. Yin. 2016. "Use of Insect Powder as a Source of Dietary Protein in Early-Weaned Piglets." *Journal of Animal Science* 94: 111–16.

Joensuu, K., and F. Silvenius. 2017. "Production of Mealworms for Human Consumption in Finland: A Preliminary Life Cycle Assessment." *Journal of Insects as Food and Feed* 3: 211–16.

Joly, G., and J. Nikiema. 2019. *Global Experiences on Waste Processing with Black Soldier Fly (Hermetia illucens): From Technology to Business.* CGIAR Research Program on Water, Land and Ecosystems, Resource Recovery and Reuse Series 16. Colombo, Sri Lanka: International Water Management Institute. doi:10.5337/2019.214.

Jongema, Y. 2017. "Worldwide List of Recorded Edible Insects." Wageningen University & Research, Netherlands. https://www.wur.nl/en/Research-Results/Chair-groups/Plant-Sciences/Laboratory-of-Entomology/Edible-insects/Worldwide-species-list.htm.

Kelemu, S., S. Niassy, B. Torto, K. Fiaboe, H. Affognon, H. Tonnang, et al. 2015. "African Edible Insects for Food and Feed: Inventory, Diversity, Commonalities and Contribution to Food Security." *Journal of Insects as Food and Feed* 1 (2): 103–19.

Kim, B. F., R. E. Santo, A. P. Scatterday, J. P. Fry, C. M. Synk, S. R. Cebron, et al. 2019. "Country-Specific Dietary Shifts to Mitigate Climate and Water Crises." *Global Environmental Change* 62: 101926. doi:10.1016/J.GLOENVCHA.2019.05.010.

Krug, J. H. A. 2017. "Adaptation of Colophospermum Mopane to Extra-Seasonal Drought Conditions: Site-Vegetation Relations in Dry-Deciduous Forests of Zambezi Region (Namibia)." *Forest Ecosystems* 4 (1).

Langley, J., S. Van der Westhuizen, G. Morland, and B. van Asch. 2020. "Mitochondrial Genomes and Polymorphic Regions of Gonimbrasia Belina and Gynanisa Maja (Lepidoptera: Saturniidae), Two Important Edible Caterpillars of Southern Africa." *International Journal of Biological Macromolecules* 144: 632–42.

Latunde-Dada, G. O., W. Yang, and A. M. Vera. 2016. "In Vitro Iron Availability from Insects and Sirloin Beef." *Journal of Agricultural and Food Chemistry* 64: 8420–24.

Lautenschläger, T., C. Neinhuis, M. Monizi, J. L. Mandombe, A. Förster, T. Henle, and M. Nuss. 2017. "Edible Insects of Northern Angola." *African Invertebrates* 58: 55–82.

Law, C. 2020. "Insect Farming: The Industry Set to Be Worth $8 Billion by 2030." *Hive Life*, October 8, 2020. https://hivelife.com/insect-farming/.

Lecocq, T. 2019. "Insects: The Disregarded Domestication Histories." In *Animal Domestication.* IntechOpen. doi:10.5772/intechopen.81834.

Li, Q., L. Zheng, N. Qiu, H. Cai, J. K. Tomberlin, and Z. Yu. 2011. "Bioconversion of Dairy Manure by Black Soldier Fly (Diptera: Stratiomyidae) for Biodiesel and Sugar Production." *Waste Management* 31: 1316–20.

Longvah, T., K. Mangthya, and P. Ramulu. 2011. "Nutrient Composition and Protein Quality Evaluation of Eri Silkworm (Samia ricinii) Prepupae and Pupae." *Food Chemistry* 128: 400–03.

Makkar, H. P. S., G. Tran, V. Heuzé, and P. Ankers. 2014. "State-of-the-Art on Use of Insects as Animal Feed." *Animal Feed Science and Technology* 197: 1–33.

MarketWatch. 2019. "Industry Research: Global Insect Feed Market Insights." News Release, MarketWatch, New York, November 5, 2019. https://www.marketwatch.com/press-release/insect-feed-market-2019-industry-price-trend-size-estimation-industry-outlook-business-growth-report-latest-research-business-analysis-and-forecast-2024-analysis-research-2019-11-05.

Mekonnen, M. M., C. M. U. Neale, C. Ray, G. E. Erickson, and A. Y. Hoekstra. 2019. "Water Productivity in Meat and Milk Production in the US from 1960 to 2016." *Environment International* 132: 1–12.

Meneguz, M., A. Schiavone, F. Gai, A. Dama, C. Lussiana, M. Renna, and L. Gasco. 2018. "Effect of Rearing Substrate on Growth Performance, Waste Reduction Efficiency and Chemical Composition of Black Soldier Fly (*Hermetia illucens*) Larvae." *Journal of the Science of Food and Agriculture* 98: 5776–84.

Mertenat, A., S. Diener, and C. Zurbrügg. 2019. "Black Soldier Fly Biowaste Treatment: Assessment of Global Warming Potential." *Waste Management* 84: 173–81.

Miglietta, P., F. De Leo, M. Ruberti, and S. Massari. 2015. "Mealworms for Food: A Water Footprint Perspective." *Water* 7: 6190–6203.

Motte, C., A. Rios, T. Lefebvre, H. Do, M. Henry, and O. Jintasataporn. 2019. "Replacing Fish Meal with Defatted Insect Meal (Yellow Mealworm Tenebrio molitor) Improves the Growth and Immunity of Pacific White Shrimp (Litopenaeus vannamei)." *Animals* 9 (5): 258.

Moula, N., and J. Detilleux. 2019. "A Meta-Analysis of the Effects of Insects in Feed on Poultry Growth Performances." *Animals* 9 (5): 201.

Muafor F. J., A. A. Gnetegha, P. Le Gall, and P. Levang. 2015. "Exploitation, Trade and Farming of Palm Weevil Grubs in Cameroon." Center for International Forestry Research, Bogor, Indonesia.

Naukkarinen, M. 2016. "Edible Insects for Improved Food and Nutrition Security at Kakuma Refugee Camp." Master's thesis, Department of Nutrition, Exercise and Sports, University of Copenhagen, Denmark.

Niassy, S., and S. Ekesi. 2017. "Eating Insects Has Long Made Sense in Africa: The World Must Catch Up." *The Conversation*, January 10, 2017. https://theconversation.com/eating-insects-has-long-made-sense-in-africa-the-world-must-catch-up-70419.

Nyakeri, E. M., M. A. Ayieko, F. A. Amimo, H. Salum, and H. J. O. Ogola. 2019. "An Optimal Feeding Strategy for Black Soldier Fly Larvae Biomass Production and Faecal Sludge Reduction." *Journal of Insects as Food and Feed* 5: 201–13.

Nyakeri, E. M. M., H. J. J. Ogola, M. A. A. Ayieko, and F. A. A. Amimo. 2016. "An Open System for Farming Black Soldier Fly Larvae as a Source of Proteins for Smallscale Poultry and Fish Production." *Journal of Insects as Food and Feed* 3: 51–56.

Nyakeri, E. M. M., H. J. J. Ogola, M. A. A. Ayieko, and F. A. A. Amimo. 2017. "Valorisation of Organic Waste Material: Growth Performance of Wild Black Soldier Fly Larvae (Hermetia illucens) Reared on Different Organic Wastes." *Journal of Insects as Food and Feed* 3: 193–202.

Oonincx, D. G. A. B., and I. J. M. de Boer. 2012. "Environmental Impact of the Production of Mealworms as a Protein Source for Humans: A Life Cycle Assessment." *PLoS One* 7: e51145.

Oonincx, D. G. A. B., S. Van Broekhoven, A. van Huis, and J. J. A. Van Loon. 2015. "Feed Conversion, Survival and Development, and Composition of Four Insect Species on Diets Composed of Food By-Products." *PLoS One* 10 (12): e0144601.

Osimani, A., V. Milanovic, F. Cardinali, A. Roncolini, C. Garofalo, F. Clementi, et al. 2018. "Bread Enriched with Cricket Powder (*Acheta domesticus*): A Technological, Microbiological and Nutritional Evaluation." *Innovative Food Science and Emerging Technologies* 48: 150–63.

Osongo, V., I. M. Osuga, C. Gachauiri, and A. M. Wachira. 2018. "Insects for Income Generation through Animal Feed: Effect of Dietary Replacement of Soybean and Fish Meal with Black Soldier Fly Meal on Broiler Growth and Economic Performance." *Journal of Economic Entomology* 111 (4): 1966–73.

Poelaert, C., F. Francis, T. Alabi, R. C. C. Megido, B. Crahay, J. Bindelle, and Y. Beckers. 2018. "Protein Value of Two Insects, Subjected to Various Heat Treatments, Using Growing Rats and the Protein Digestibility-Corrected Amino Acid Score." *Journal of Insects as Food and Feed* 4: 77–87.

Pomalégni, S. C. B., D. S. J. C. Gbemavo, C. P. Kpadé, M. Kenis, and G. A. Mensah. 2018. "Traditional Poultry Farmers' Willingness to Pay for Using Fly Larvae Meal as Protein Source to Feed Local Chickens in Benin." *Bio-Based and Applied Economics* 7: 117–38.

RDA (Rural Development Administration). 2020a. "Development Policy and Plans for Korean Insect Industry." RDA, Wansan-gu, Jeonju, Republic of Korea.

RDA (Rural Development Administration). 2020b. "Survey of Current Status of Korean Insect Industry, 2020." RDA, Wansan-gu, Jeonju, Republic of Korea.

Renna, M., A. Schiavone, F. Gai, S. Dabbou, C. Lussiana, V. Malfatto, et al. 2017. "Evaluation of the Suitability of a Partially Defatted Black Soldier Fly (Hermetia illucens L.) Larvae Meal as Ingredient for Rainbow Trout (Oncorhynchus mykiss Walbaum) Diets." *Journal of Animal Science and Biotechnology* 8: 57.

Reuters. 2020. "Astanor Raises $325 Million Fund to Invest in Agri-Food Tech Startups." Reuters, November 20, 2020. https://www.reuters.com/article/us-tech-astanor/astanor-raises-325-million-fund-to-invest-in-agri-food-tech-startups-idUSKBN2800M6.

Robinson, T., and F. Pozzi. 2011. "Mapping Supply and Demand for Animal-Source Foods to 2030." Animal Production and Health Working Paper No. 2, Food and Agriculture Organization of the United Nations, Rome.

Roffeis, M., E. C. Fitches, M. E. Wakefield, J. Almeida, T. R. Alves Valada, E. Devic, et al. 2018. "Life Cycle Cost Assessment of Insect Based Feed Production in West Africa." *Journal of Cleaner Production* 199: 792–806.

Roffeis, M., E. C. Fitches, M. E. Wakefield, J. Almeida, T. R. Alves Valada, E. Devic, et al. 2020. "Ex-ante Life Cycle Impact Assessment of Insect Based Feed Production in West Africa." *Agricultural Systems* 178: 1–21.

Roos, N. 2018. "Insects and Human Nutrition." In *Edible Insects in Sustainable Food Systems*, edited by A. Halloran, R. Flore, P. Vantomme, and N. Roos, 363–74. Springer International Publishing. doi:10.1007/978-3-319-74011-9_5.

Rumpold, B. A., and O. K. Schluter. 2013. "Nutritional Composition and Safety Aspects of Edible Insects." *Molecular Nutrition & Food Research* 57: 802–23.

Sánchez-Bayo, F., and K. A. G. Wyckhuys. 2019. "Worldwide Decline of the Entomofauna: A Review of Its Drivers." *Biological Conservation* 232: 8–27.

Selenius, O. V. O., J. Korpela, S. Salminen, and C. G. Gallego. 2018. "Effect of Chitin and Chitooligosaccharide on In Vitro Growth of Lactobacillus rhamnosus GG and Escherichia coli TG." *Applied Food Biotechnology* 5: 163–72.

SG Links. 2019. "Written Reply by Masagos Zulkifli, Minister for the Environment and Water Resources, to Parliamentary Question on Insect Farms on 7 October 2019." Ministry of Environment and Water Resources, Singapore. http://sglinks.news/mewr/pr/written-reply-masagos-zulkifli-minister-environment-water-resources-17cdb3b.

Skau, J. K. H., B. Touch, C. Chhoun, M. Chea, U. S. Unni, J. Makurat, et al. 2015. "Effects of Animal Source Food and Micronutrient Fortification in Complementary Food Products on Body Composition, Iron Status, and Linear Growth: A Randomized Trial in Cambodia." *American Journal of Clinical Nutrition* 101: 742–51.

Smetana, S., M. Palanisamy, A. Mathys, and V. Heinz. 2016. "Sustainability of Insect Use for Feed and Food: Life Cycle Assessment Perspective." *Journal of Cleaner Production* 137: 741–51.

Smith, M. R., V. J. Stull, J. A. Patz, and S. S. Myers. 2021. "Nutritional and Environmental Benefits of Increasing Insect Consumption in Africa and Asia." *Environmental Research Letters* 16.

Sogari, G., M. Amato, I. Biasato, S. Chiesa, and L. Gasco. 2019. "The Potential Role of Insects as Feed: A Multi-Perspective Review." *Animals* 9 (4).

Ssepuuya, G., C. Sebatta, E. Sikahwa, P. Fuuna, M. Sengendo, J. Mugisha, K. K. M. Fiaboe, and D. Nakimbugwe. 2019. "Perception and Awareness of Insects as an Alternative Protein Source among Fish Farmers and Fish Feed Traders." *Journal of Insects as Food and Feed* 5: 107–16.

St-Hilaire, S., K. Cranfill, M. A. McGuire, E. E. Mosley, J. K. Tomberlin, L. Newton, et al. 2007. "Fish Offal Recycling by the Black Soldier Fly Produces a Foodstuff High in Omega-3 Fatty Acids." *Journal of the World Aquaculture Society* 38: 309–13.

Stull, V. J., E. Finer, R. S. Bergmans, H. P. Febvre, C. Longhurst, D. K. Manter, et al. 2018. "Impact of Edible Cricket Consumption on Gut Microbiota in Healthy Adults, a Double-Blind, Randomized Crossover Trial." *Scientific Reports* 8: 10762.

Tanga, C. M., H. J. O. Magara, M. A. Ayieko, R. S. Copeland, F. M. Khamis, S. A. Mohamed, F. O. Umbura, S. Niassy, S. Subramanian, K. K. M. Foaboe, N. Roos, S. Ekesi, and S. Huge. 2018. "A New Edible Cricket Species from Africa of the Genus Scapsipedus." *Zootaxa* 4486 (3): 383–92. https://doi.org/10.11646/zootaxa.4486.3.9.

Temple, W. D., R. Radley, J. Baker-French, and F. Richardson. 2013. "Use of Enterra Natural Fertilizer (Black Soldier Fly Larvae Digestate) as a Soil Amendment." Project Report, Enterra Feed Corporation, Vancouver, Canada.

Terova, G., S. Rimoldi, C. Ascione, E. Gini, C. Ceccotti, and L. Gasco. 2019. "Rainbow Trout (Oncorhynchus mykiss) Gut Microbiota Is Modulated by Insect Meal from Hermetia illucens Prepupae in the Diet." *Reviews in Fish Biology and Fisheries* 29: 465–86.

Tinder, A. C. C., R. T. T. Puckett, N. D. D. Turner, J. A. A. Cammack, and J. K. K. Tomberlin. 2017. "Bioconversion of Sorghum and Cowpea by Black Soldier Fly (Hermetia illucens (L.)) Larvae for Alternative Protein Production." *Journal of Insects as Food and Feed* 3: 121–30.

TNO (Organisation for Applied Scientific Research). 2019. "Short Feasibility of Flying Food in Burundi." TNO, The Hague, Netherlands. https://www.tno.nl/en/about-tno/news/2019/11/producing-crickets-for-food-is-viable-in-burundi/.

UNCTAD (United Nations Conference on Trade and Development). 2018. "90% of Fish Stocks Are Used Up—Fisheries Subsidies Must Stop." UNCTAD, Geneva. https://unctad.org/en/pages/newsdetails.aspx?OriginalVersionID=1812.

van Huis, A. 2019. "Insects as Food and Feed, a New Emerging Agricultural Sector: A Review." *Journal of Insects as Food and Feed* 6 (1): 27–44.

van Huis, A., and D. G. A. B. Oonincx. 2017. "The Environmental Sustainability of Insects as Food and Feed: A Review." *Agronomy for Sustainable Development* 37.

van Huis, A., J. Van Itterbeeck, H. Klunder, E. Mertens, A. Halloran, G. Muir, and P. Vantomme. 2013. "Edible Insects: Future Prospects for Food and Feed Security." Food and Agriculture Organization of the United Nations, Rome.

van Zanten, H. H. E., P. Bikker, B. G. Meerburg, and L. J. M. de Boer. 2018. "Attributional versus Consequential Life Cycle Assessment and Feed Optimization: Alternative Protein Sources in Pig Diets." *International Journal of Life Cycle Assessment* 23: 1–11.

Vernooij, A. G., and T. Veldkamp. 2019. *Insects for Africa: Developing Business Opportunities for Insects in Animal Feed in Eastern Africa*. Netherlands: Wageningen Research.

Wang, L. F., and G. Crameri. 2014. "Emerging Zoonotic Viral Diseases." *OIE Revue Scientifique et Technique* 33: 569–81.

Watanabe, F., and T. Bito. 2018. "Vitamin B12 Sources and Microbial Interaction." *Experimental Biology and Medicine* 243: 148–58.

Wilbur-Ellis. 2018. "New COO and NSF Grant Fuel Beta Hatch Growth and Efficiency." Wilbur-Ellis Company, San Francisco. https://www.wilburellis.com/new-coo-and-nsf-grant-fuel-beta-hatch-growth-and-efficiency/.

Willett, W., J. Rockström, B. Loken, M. Springmann, T. Lang, S. Vermeulen, et al. 2019. "Food in the Anthropocene: The EAT–Lancet Commission on Healthy Diets from Sustainable Food Systems." *Lancet* 393: 447–92.

Yen, A. L. 2009. "Entomophagy and Insect Conservation: Some Thoughts for Digestion." *Journal of Insect Conservation* 13: 667–70.

Yen, A. L. 2015. "Insects as Food and Feed in the Asia Pacific Region: Current Perspectives and Future Directions." *Journal of Insects as Food and Feed* 1: 33–55.

Ynsect. 2019. "Media Coverage." Ynsect, Evry, France. http://www.ynsect.com/wp-content/uploads/2019/03/Press-review-Ynsect-1.pdf.

Yu, M., Z. Li, W. Chen, T. Rong, G. Wang, J. Li, and X. Ma. 2019. "Use of Hermetia illucens Larvae as a Dietary Protein Source: Effects on Growth Performance, Carcass Traits, and Meat Quality in Finishing Pigs." *Meat Science* 158.

CHAPTER FOUR

Mainstreaming Insect Farming

HIGHLIGHTS

- Insect supply chains in Africa are largely informal and differ by insect species and country, but they are slowly becoming an established part of the food system.
- Insect farming costs are determined by labor needs, transport costs, road accessibility, substrate availability, infrastructure requirements, and fuel and electricity requirements.
- Small-scale insect farming requires minimal infrastructure, does not require climate control, and may require only an open pen or crate.
- Small-scale insect farming is low-tech and requires a limited amount of labor and processing to add value to farmed insects.
- Specialized and commercial insect production systems are more cost-effective than integrated, or generalized, systems.
- Modeling of potential widespread black soldier fly (BSF) farming in Africa shows that Africa's agricultural waste could supply about 200 million tons of substrate.
- With this substrate, black soldier fly larvae (BFSL) could produce enough protein meal to meet up to 14 percent of the crude protein needs to rear all the pigs, goats, fish, and chickens in Africa.
- The modeling shows that the continent could use BSFL to replace 60 million tons of traditional feed production and increase Africa's production of organic fertilizer by 60 million tons.

- Establishing a BSFL industry in Africa could create 15 million jobs.
- The industry would prevent 86 million tons of carbon dioxide equivalent (CO_2-eq) emissions, which is the equivalent of removing 18 million vehicles from the roads.
- In all of Africa, BSF farming could produce a market value of crude protein worth up to US$2.6 billion and fertilizer worth up to US$19.4 billion.

The purpose of this chapter is to understand the required processes for mainstreaming insect farming into a circular food economy. Mainstreaming insects can be understood in two distinct ways. First, it could refer to establishing commercially viable insect production systems, like those in the Republic of Korea and some other countries. Second, it could refer to cultural acceptance and the integration of insects into dietary practices, as is happening in many African countries, including those affected by fragility, conflict, and violence (FCV). The first section describes the general supply chains of edible insects, finding that they differ among insects and countries and are often informal. The second section looks at edible insect markets and finds key differences between urban and rural markets. The third section examines the factors that drive the market and establish costs. It finds that these factors differ for rural, small-scale operations and larger, commercial operations. The fourth section examines the value chains for crickets and BSF and the specific production systems of six types of insects, including houseflies, crickets, mealworms, silkworm chrysalids, palm weevil larvae, and BSF, finding value in the production of each insect. The fifth section models the potential mainstreaming of BSF production in Zimbabwe and other African countries by calculating specific social, economic, and environmental benefits.

EDIBLE INSECT SUPPLY CHAINS IN AFRICAN FCV-AFFECTED STATES

Insect supply chains are largely informal in Africa, but they are slowly becoming an established part of the food system. Informal markets and supply chains for wild insects have been a common part of the food supply chain in many African countries, including FCV countries. With the rise in urbanization, traditional supply chains for edible insects are becoming slightly more formalized as traders form relationships with rural farmers and collectors to meet the growing demand for selected edible insects in urban markets. In Zimbabwe, for example, a relatively robust supply chain for wild insects with differentiated stakeholders has evolved (figure 4.1).

The supply chain for wild harvested mopane caterpillars in Zimbabwe is an example of the types of distribution channels through which farmers sell their harvests. These include the following four distribution channels. (1) Collectors sell to local shopkeepers who sell the insects directly to local consumers or transport and sell the insects to urban market retailers.[1] (2) A family member

FIGURE 4.1 Zimbabwe's Wild Harvested Mopane Caterpillar Supply Chain

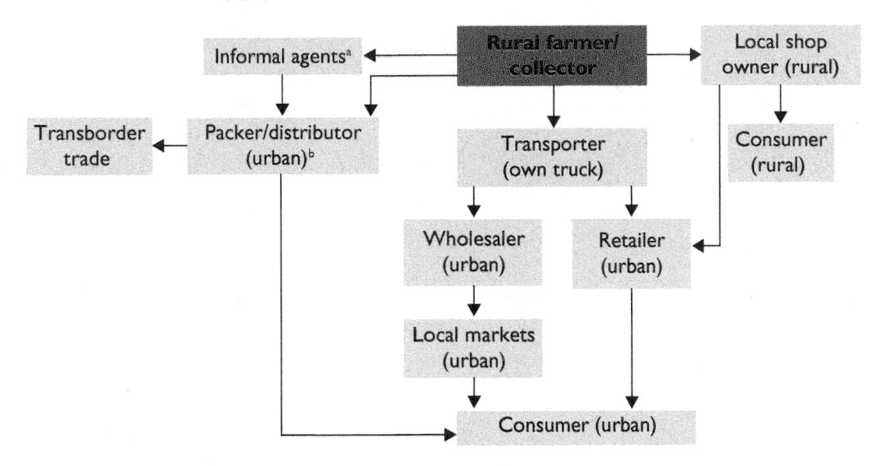

Source: Original figure for this publication.
a. Family and friends, sales commission.
b. Distribute other agricultural products.

or friend acts as the collector's informal agent to sell the mopane caterpillars for commission to an urban packer or distributor.[2] (3) The collector sells the insects directly to a packer or distributor, who in turn sells the insects to consumers in urban markets through existing agricultural product distribution channels. Alternatively, if packers or distributors have existing cross-border distribution channels, they may leverage the channels to export the caterpillars. (4) The collector sells to a transporter who collects the caterpillars from the collector's village and has distribution arrangements with urban wholesalers and retailers. For this fourth channel, the collector must consolidate enough insects for the transporter to cover the costs of travel.

The supply chain structure for edible insects in other countries is different from the wild harvested mopane caterpillar supply chain in Zimbabwe. In the Democratic Republic of Congo, peak insect season harvests are bountiful, causing insect supply surpluses (figure 4.2). As a result, wholesalers of wild caught insects in the Democratic Republic of Congo store dried insects for short periods to normalize the supply-demand imbalance. This allows wholesalers to manage transborder trade with regional markets. In non-African FCV-affected states, other unique supply chains exist. In Papua New Guinea, rural farmers and collectors share their wild insect harvests, free of cost, directly with local communities or their immediate tribal group. The farmers sell to urban wholesalers only when there is sufficient surplus. The wholesalers then sell the surplus to local vendors in urban markets (figure 4.3).

FIGURE 4.2 Democratic Republic of Congo's Wild Harvested Edible Insect Supply Chain

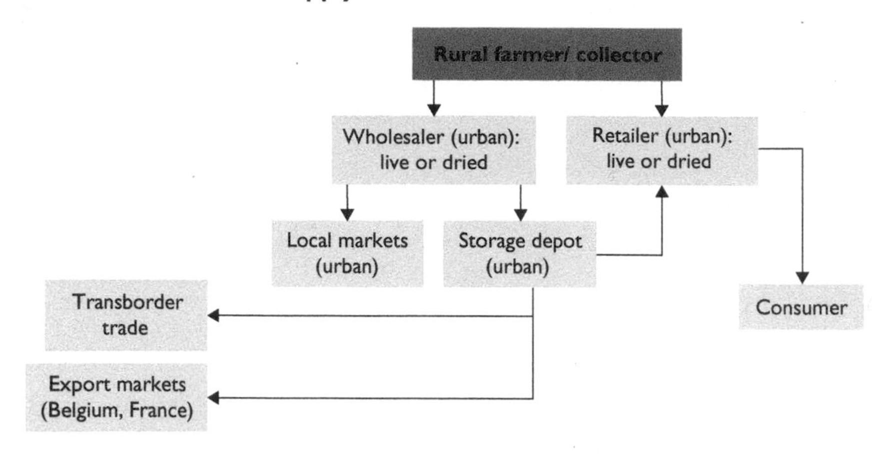

Source: Original figure for this publication.

FIGURE 4.3 Papua New Guinea's Wild Harvested Edible Insect Supply Chain

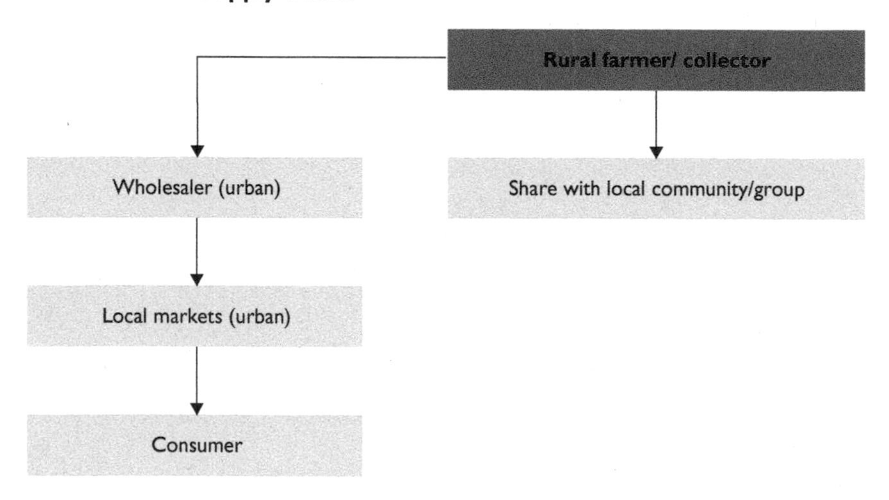

Source: Original figure for this publication.

URBAN AND RURAL INSECT MARKETS

Insect consumption varies between urban and rural areas. In Zimbabwe, for example, a nonrandomized survey indicated that consumption of wild harvested insects is three times more common in rural than urban areas (Manditsera et al. 2018). According to the survey, "taste" is the main motive for people to eat insects. Urban dwellers, however, said they are more motivated to consume insects because of the insects' nutritional value and medicinal properties. Greater availability of edible insects was related to greater insect consumption in both urban and rural areas.

There are no commonly shared factors determining insect consumption within Africa's rural areas. Even rural Sub-Saharan African ethnic groups living within geographic proximity to each other exhibit different preferences for wild harvested insects and consider consuming certain species taboo (Kelemu et al. 2015). In some cases, insect consumption is partly driven by economic and dietary necessities. In Zimbabwe, for example, there is a clear pattern that when there is a lean period, or a period with low harvests and food shortages, insect availability and consumption increase (see table 4.1 for details on the insect harvest and consumption schedule).

African urban markets have nuanced demands for edible insects. In Zambia, for example, certain types of wild harvested insects—such as *Coleoptera* (beetle), *Hemiptera* (cicadas), and *Hymenoptera* (edible bee larvae)—are less desirable in urban markets because they are consumed by the "lower class and physical laborers" (Stull et al. 2018). By contrast, certain insects—such as *Isoptera* (termites, locally referred to as *inswa*) and *Lepidoptera* (caterpillars, locally referred to as *vinkubala*)—are considered a "delicacy" and consumed by the "upper class" and white-collar workers. As a result, these insects are in higher demand and command higher prices in urban markets (figure 4.4).

TABLE 4.1 Calendar for Crop and Wild Insect Harvesting in Zimbabwe and Description of When Consumption Occurs

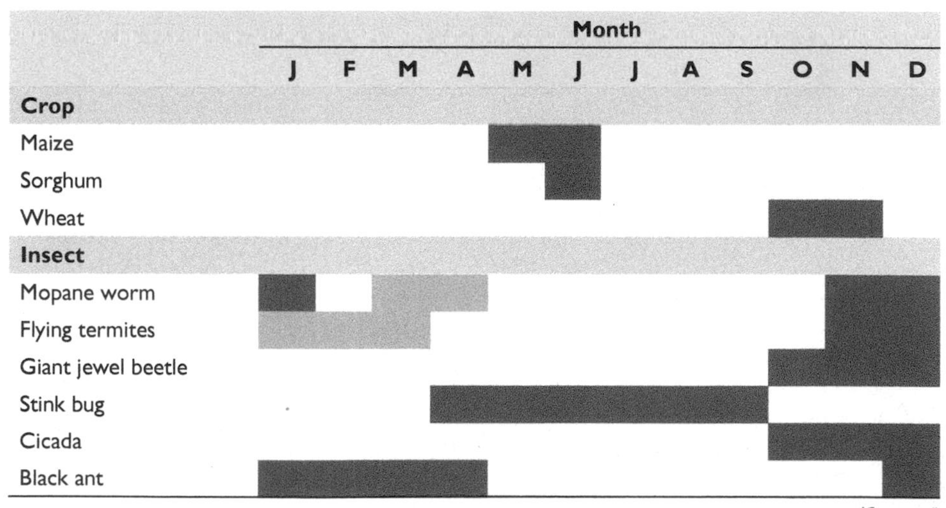

	Month											
	J	F	M	A	M	J	J	A	S	O	N	D
Crop												
Maize												
Sorghum												
Wheat												
Insect												
Mopane worm												
Flying termites												
Giant jewel beetle												
Stink bug												
Cicada												
Black ant												

(Continued)

TABLE 4.1 Calendar for Crop and Wild Insect Harvesting in Zimbabwe and Description of When Consumption Occurs *(Continued)*

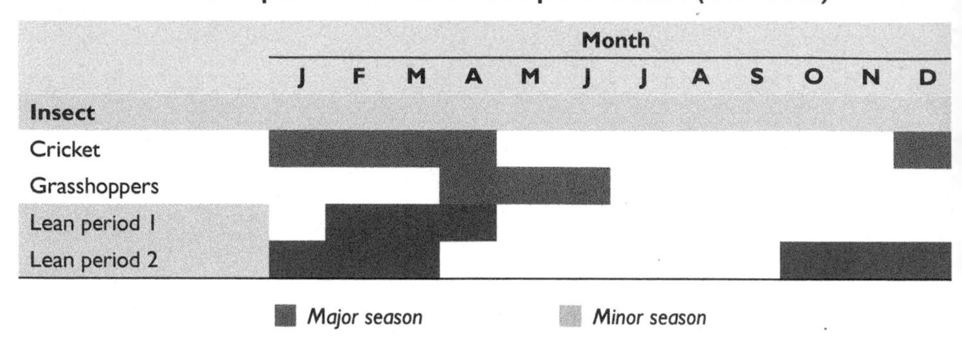

Insect	J	F	M	A	M	J	J	A	S	O	N	D
Cricket	■	■	■	■								■
Grasshoppers				■	■	■						
Lean period 1		■	■	■								
Lean period 2	■	■	■							■	■	■

■ *Major season* ■ *Minor season*

Mopane: 10 percent consumed immediately after harvest; 90 percent dried to be consumed postharvest several months later.
Flying termites: 90 percent consumed within two weeks of harvesting and 10 percent within a month. Termites are fatty and easily go rancid; thus, they cannot be stored for a long time beyond two months.
Zvigakata: 50 percent consumed immediately after harvest; 50 percent stored for long-term consumption, within three months of harvest.
Stink bug: 90 percent consumed within the first few days of harvest; 10 percent stored and consumed within three weeks of harvest. They cannot be stored for long due to high fat content.
Nyeza: Not widely consumed due to difficulty of harvest; 100 percent consumed soon after harvest; they have a short shelf life.
Tsambarafuta: 80 percent consumed immediately after harvest, 20 percent within a month after harvest. It is very difficult to gather large quantities from the wild.
Cricket: 60 percent consumed immediately after harvest; 40 percent dried, salted, and consumed within a month of harvest.
Grasshoppers: 50 percent consumed immediately after harvest; 50 percent salted and dried and consumed within three months of harvest.
Source: Original figure for this publication, using information from Robert Musandire, Chinhoyi University of Technology, Zimbabwe.

Two parallel market structures for farmed insects are likely to evolve in urban and rural areas. These structures include a traditional rural market close to the points of harvest and, as commercial insect production expands and introduces predictability in the market, an urban market with wholesale-retail supply chains catering to urban consumers. Commercial insect producers will most likely supply livestock feed and insect-based fertilizers to urban and peri-urban areas, where most large-scale, commercial agricultural producers, processors, and markets reside. However, commercial insect production in urban markets may or may not affect insect consumption habits in rural

FIGURE 4.4 Nontribal Social Arrangements of Wild Harvested Edible Insects in Zambia's Kazoka Village

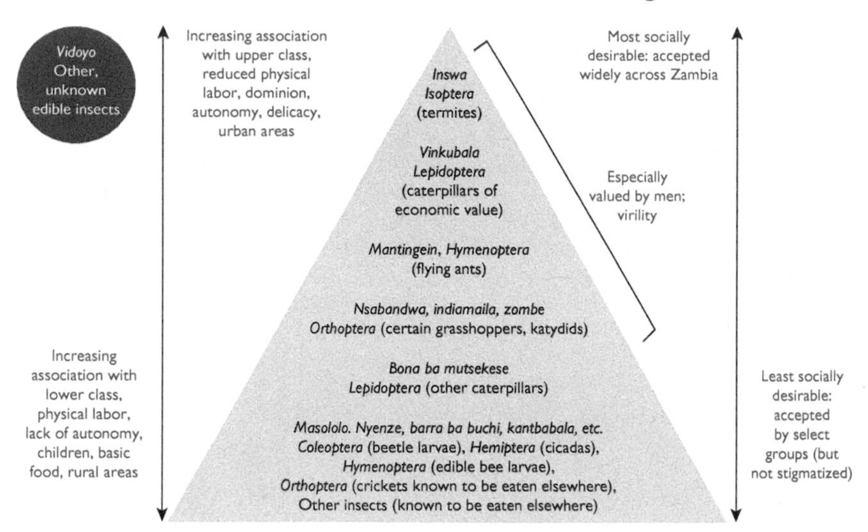

Source: Stull et al. 2018.

areas, where wild insect harvesting will likely continue. One possible change to rural markets may include the development of simple, small-scale production systems—at the household or community level. These systems could ensure more predictable supplies of edible insects for rural markets and reduce the risk of overharvesting.

DRIVERS OF THE EDIBLE INSECT MARKET

The development of insect-related production systems and their costs are determined by several factors. Figure 4.5 depicts a high-level overview of the general supply and value chains in the insect and insect-based product industry. The figure also outlines the key factors within the enabling environment that influence the evolution of these supply and value chains. These key factors include breeding requirements; product development factors; risk management issues related to spoilage, allergens, storage and processing impacts, and chemical or environmental hazards; environmental impact such as greenhouse gas (GHG) emissions, energy exploitation, water and natural resource stress, and land and forest degradation; industry regulations; bioethics; behavioral changes related to insect consumption; and existing or lacking distribution channels. Insect-related production systems include small-scale rural systems and commercial systems and the associated markets and supply chains that link them. The most

FIGURE 4.5 Rough Representation of the Farmed Edible Insect Value and Supply Chains

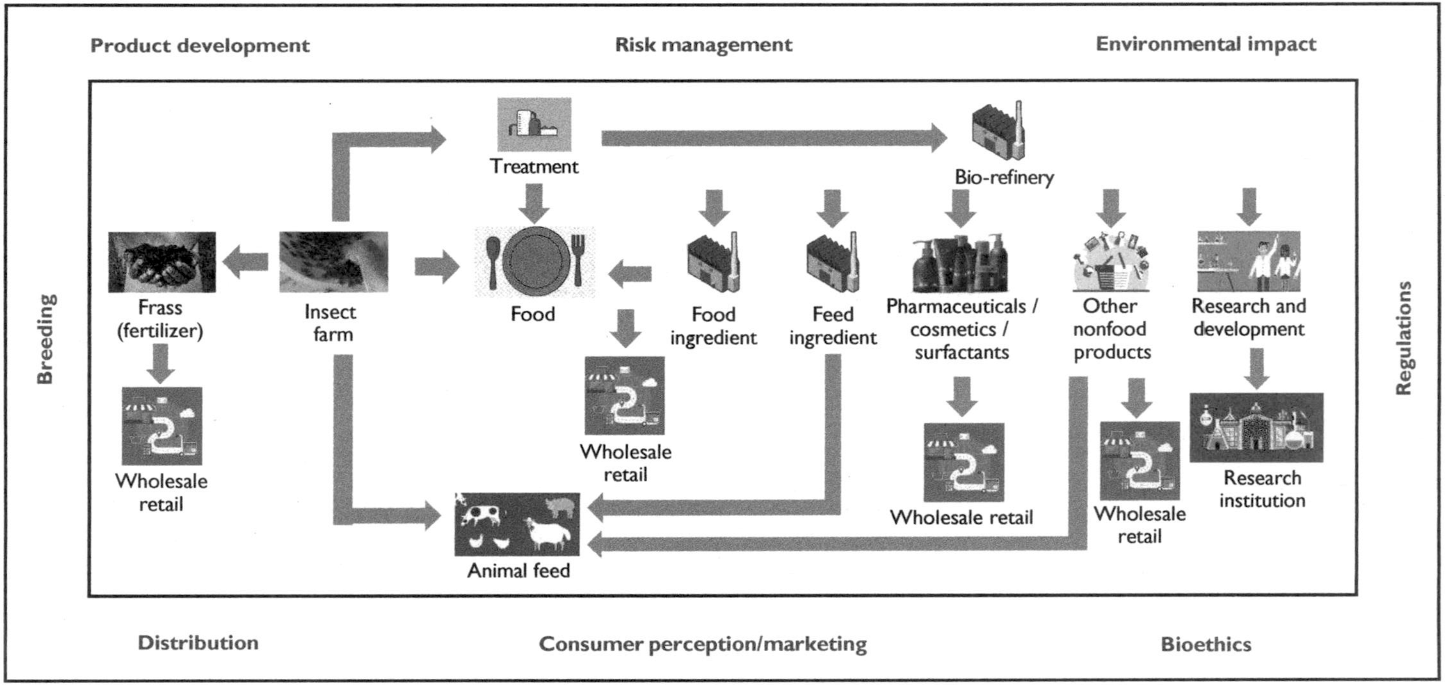

Source: Adaptation of Peters, undated.

TABLE 4.2 Key Factors Associated with the Costs of Small-Scale and Commercial Insect Production Systems

Small-scale, rural production	Commercial production
• Substrate availability	• Substrate availability
• Labor cost	• Quality of transport infrastructure
• Infrastructure costs	• Transport costs
• Quality and accessibility of roads and other transport infrastructure	• Fuel costs
• Transport costs	• Cost, availability, and quality of electricity
• Fuel costs	• Labor cost

Source: Original table for this publication.

significant of these factors are identified in table 4.2 and described in the following subsections.

Large-scale, commercial insect farming can exist in parallel with small-scale insect farming. For example, some small-scale farmers can farm for their own home or farm use or sell in local markets, while others could join producer groups, including as subcontractors, that act as commercial enterprises and compete with larger producers in broader markets. Some farmers may even choose to do both at the same time. The research team found examples of this with cricket farming in Thailand and mealworm farming in Korea. In other cases, a large commercial BSFL processor may outsource production to small farms in rural areas. Either way, this is not taking over a small-scale operation—it is outsourcing. When large-scale producers move into areas of more artisanal producers, the small-scale farmers may lose competitive access to local markets but could still produce to meet their own home or on-farm livestock needs. That said, even within the same country or local area, there may be different markets for farmed insects to which both large-scale and small-scale producers can cater. In Nairobi, Kenya, for example, there are farms around the capital that cater to BSF markets since these areas have access to food processors, waste resources, and feed companies. In rural Kenya, by contrast, small-scale cricket farming is more common as it can cater to the local market.

Small-Scale Rural Insect Production

Substrate Availability

Most existing insect farms are small-scale with costs determined by substrate availability. Of the farms reached by the farm-level survey, 76 percent are characterized as "small-scale," 20 percent are "medium-scale," and 4 percent are "large-scale."[3] This is aligned with the World Bank team's observation from field visits that the African insect sector comprises many small producers. A

Kenyan insect farmer described this: "The crickets consume very little water and food and require a small amount of space for rearing and hence I can do it in my house." For insects that require particular substrates to feed on, certain locations may have comparative advantages if the substrate is locally available. However, this is not true for many insects—such as crickets, BSF, and the common housefly—that consume a broader range of substrates, offering flexibility to small-scale producers.

Technology and Other Infrastructure

Small-scale insect farming infrastructure is limited in size, does not require climate control, and may require only an open pen or crate. The basic infrastructure needed for insect farming is an insect containment structure, which varies in size and sophistication depending on the farmer's needs, the production system's scale, and the insect species being farmed. The most common types of containment structures are a crate system (20 percent), in which containers are vertically stacked on top of one another, and open pens (30 percent), which are not stacked. Forty-seven percent of farms use some other kind of containment structure and 3 percent use multiple structures. According to the farm-level survey, only 4 percent of farms are fully contained and climate controlled, while 53 percent are without climate control. Another 39 percent of farms are operated under open air, while 4 percent use "another" form of housing. So far, small farms have not incorporated digital technologies.

Labor

Small-scale insect farming is low-tech and requires a limited amount of labor and processing to add value to farmed insects. According to the farm-level survey, insect farms in Africa tend to be low-tech and operated almost entirely by manual labor (97 percent). Only 3 percent of the surveyed farms had automated components, and none was fully automated. The survey found very few farms that process the insects, with only 6 percent drying the insects, 2 percent boiling them, and 1 percent grinding them. The other 91 percent of farms sell fresh, unprocessed insects. This limited processing reduces labor requirements. Several African farmers repeated this point. A male farmer from Rwanda said, "[Insect] farming requires much less work than other types of farming." A female farmer from Madagascar said, "There is no need for male labor, I can do all the activities myself," while another said, "Insect farming covers the needs of my family while leaving time to do other social responsibilities and obligations."

Commercial Insect Production

Local Substrate Availability

There are substantial cost implications if substrates must be imported, particularly for countries without internal resources. For example, if

Zimbabwe—a landlocked country with limited local substrate resources—sourced substrates from overseas, the shipment would have to transit through Durban, South Africa. The cost of transporting goods from Durban to Harare is US$0.14/kilometer-ton. This high transportation cost would account for 55 to 65 percent of the substrate's value, making import costs prohibitively high. There are some local substrate options, but these are often used for other industries. For example, brewery waste can be used as a locally sourced substrate for BSFL. However, the livestock industry uses this waste to produce livestock feed. The number and variety of potential waste products is reduced if insect producers plan to export insect products to the European Union or the United States, where strict substrate requirements are in place. Even so, most insects have particular dietary needs and so rely on limited types of substrates. For example, the mopane worm feeds only on mopane tree leaves.

Connectivity

Poor-quality roads and low connectivity from producers to consumers can slow the advancement of any industry, including the farmed insect industry. Poor-quality roads and transport infrastructure are chronic problems throughout much of Africa, particularly in African FCV countries. Infrastructure inefficiencies and their impacts vary among Africa's regions. However, according to a World Bank study, rehabilitating roads in East Africa's transport corridor from fair to good would reduce transport costs by 15 percent and transport prices by up to 10 percent (Teravaninthorn and Raballand 2009). In Southern Africa's transport corridor, rehabilitation would reduce costs by up to 5 percent and prices by up to 3 percent. In corridors in West and Central Africa, rehabilitation would reduce costs by 5 percent but, because of strongly regulated markets, would not affect transport prices. Thus, the development of insect supply chains and market linkages will be defined by the costs associated with reaching potential markets. In many African FCV countries, the supply chain and market reach of commercial insect producers are expected to be limited. In Zimbabwe, for example, 68 percent of the population lives in rural areas, which are also where most insects and insect-based products are consumed. Given the poor quality of road and transport infrastructure in these areas, certain analyses could inform commercial insect producers about which factors are under their control and define the economic viability of selling insects to rural markets.

Transport and Fuel

Transportation and fuel costs will determine, in part, how supply chains for commercially produced insects and insect products will develop. In most African FCV countries, the availability and quality of transport logistics support are limited and expensive. This is because of the lack of transport options, which diminishes competition; high operating costs associated with the lack of spare parts to repair and maintain vehicles; high fuel costs, particularly in landlocked states; and poor road conditions. The World Bank compared trucking costs in the European Union and several other countries against those in four African trade corridors. The average cost per kilometer-ton was 2 cents in Pakistan, 3.5 cents in Brazil, 5 cents in the United States and China, and 7 cents in Europe. By contrast, the average costs per kilometer-ton in the four African transport corridors were 6 cents in the Durban-Lusaka corridor, 7 cents in Lomé-Ouagadougou, 8 cents in Mombasa-Kampala, and 11 cents in Douala-N'Djamena (Teravaninthorn and Raballand 2009). Costs associated with transporting substrates to production facilities are an important consideration when developing a production system. These costs include the cost of transport and the acquisition cost of substrates. This is particularly true if the insects bred have specific dietary needs. For example, commercially producing mopane caterpillars may not be economically feasible if mopane trees are not within a reasonable proximity.

Energy

Commercial insect and insect product production requires a reliable energy source. This is necessary for lighting and climate control systems and processing, such as drying or milling. In many African FCV countries, expensive and unreliable electricity is an ongoing challenge, making backup generators necessary to maintain uninterrupted production. However, fuel-generated electricity costs are approximately three times that of on-grid electricity. In Zimbabwe, for example, the cost of on-grid electricity for a commercial operation is approximately US$0.05/kilowatt-hour compared with US$0.15/kilowatt-hour from a diesel generator. A commercial producer would also need to account for the cost of generator repairs and maintenance, which can be costly for imported generators with foreign parts. Renewable energy sources may provide economically viable alternatives to fuel-powered generators for off-grid electricity generation given the relatively low energy requirements, predominantly for lighting, for small-scale insect breeding. Agricultural systems are increasingly using solar photovoltaics (PV), or solar panels. Solar PV is competitive for off-grid or mini-grid applications where the main alternative is diesel or gasoline generators (IEA 2014). A small rooftop panel could generate enough electricity to power the few energy-efficient light bulbs necessary for insects. Micro-hydro power generation may be even less capital intensive than solar PV if the insect breeding system is situated near a source of consistently flowing water, like a river. Hydropower also has the advantage of being a 24-hour, 365-day power source, assuming the water flow is not seasonal (Sims et al. 2015).

Labor

Industrial-scale production of edible insects and insect products can be labor intensive and requires new skills. For example, 75 percent of cricket farming costs are for labor (Lynley 2018). Animal protein production was optimized long ago, but the insect sector has not benefited from similar advances. Insect farming is a nascent field, so the technology is still developing. As such, many commercial producers hope to transition to automated production systems to produce insects efficiently without an expensive labor force. Moreover, the edible insect industry's labor force requires new skills. According to the International Platform of Insects for Food and Feed, insect production requires knowledge of (1) insect behavior, (2) food and feed safety principles, (3) insect species, (4) farmed insect life cycles, and (5) insect handling and measures to prevent them from escaping (IPIFF 2019). In addition to these insect farming skills, a wide range of other skills are related to edible insects. For example, culinary academies, chefs in training, and other members of the hospitality industry must learn how to use insects to prepare tasty and nutritious meals. A study of aspiring chefs found that 86 percent were moderately or highly likely to use insects for dishes after undergoing a four-hour informational and tasting session (Halloran and Flore 2018).

Costs

Specialized insect production systems are more cost-effective than integrated, or generalized, systems. In a fully integrated insect production system, a single producer carries out all the steps in the insect production process. For example, the producer is responsible for acquiring substrates, producing and processing the insects, and marketing the final product. In a specialized system, specialists carry out the steps separately. For example, the producer is responsible for rearing insects, but a third party would acquire substrates. Commercial production systems are likely to be specialized systems, whereas small-scale, rural production systems are likely to be fully integrated. The cost per unit of production in integrated systems is expected to be higher than the cost per unit in a specialized system. These costs are largely determined by the system's access to capital, information, and technology and equipment. In rural areas, particularly in African FCV countries, access to these inputs is limited; therefore, the unit cost of production is expected to remain high until the industry grows and matures, compared with large-scale, commercial production systems, which are expected to have access to all the factors of production and marketing. Figure 4.6 shows how unit costs fall as a production system becomes more specialized.

Small-scale, manually managed insect farms require minimal infrastructure. The main investment is for the insect containment structure. These can be simple like those of rural cricket farms in Thailand and, more recently, Kenya; farmers can easily contain crickets by surrounding them with tile and plastic walls or other slippery barriers that the crickets cannot pass. The farm infrastructure—which includes pens, roofs, and walls—can be a major investment. These costs depend on labor expenses and the types of materials used for the structures. For example, the crate used to contain BSFL can be any shallow container that is easy to handle and clean and stacks on other containers to save space. Larger-scale farms require greater investment to

FIGURE 4.6 Supply Chain Integration versus Costs over Time

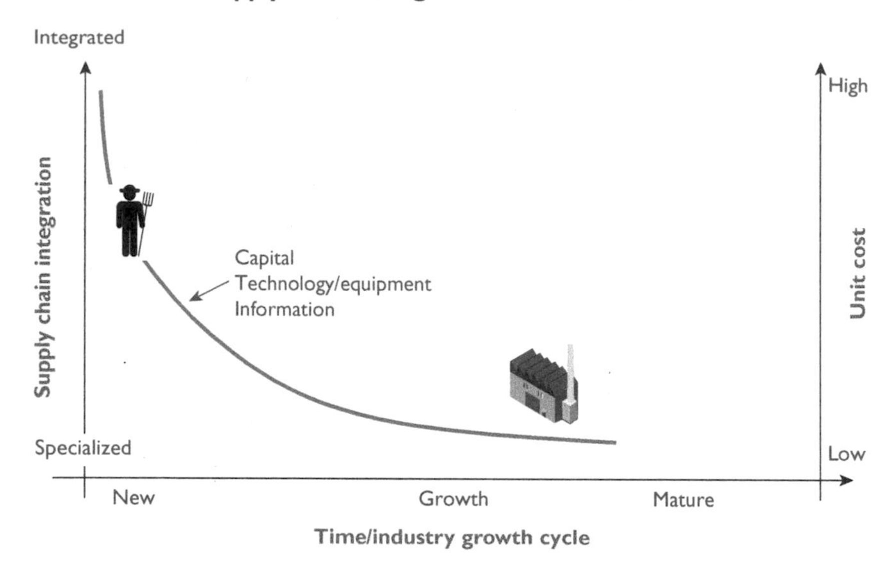

Source: Original figure for this publication.

protect the farms from predators. For example, cricket farms protect open-air cricket pens with nets to keep out birds, rodents, and lizards. Box 4.1 describes in more detail the types of inputs that are needed to produce crickets and the costs. The initial cost of producing 5 kilograms (kg) of crickets is US$15.62 in Kenya (table 4.3). The cost per kilogram of produced insects would be less in a fully commercialized system. An Egyptian farmer stated this succinctly when discussing insect-based protein powder: "Scalability is crucial for this industry because it is easier to sell 10,000 tons of protein powder than 100 tons."

Commercial producers are likely to bring down production costs quickly to crowd out competitors. As commercial producers gain access to capital, information, and technology, they are likely to flood the market with low-cost insects and insect products. At least initially, these products will be focused on insect-based protein for human and animal consumption and frass for nutrient-rich organic fertilizer. However, in African FCV countries, poor-quality roads and transport infrastructure and high fuel and transport costs could limit commercial producers' market reach. Thus, small-scale, rural producers could be insulated from the entry of low-cost insects and insect products in rural markets, despite higher unit production costs and, consequently, higher insect prices in rural areas.

As the insect industry in Africa matures, the market will introduce more standards and differentiate insect products according to consumer preferences.

BOX 4.1 Costs Associated with an Experimental Cricket Farming Activity in Kenya's Kakuma Refugee Camp

Cricket farms provide a good example of what types of infrastructure investments are needed. The experimental cricket farming activity in Kenya's Kakuma refugee camp provides cricket farmers a rearing kit consisting of six months' worth of substrate, cotton wool and disposable plates for egg laying, egg incubation boxes, carrier bags to carry the supplies, cardboard egg trays where the crickets can hide, a water sprayer to provide water to the crickets, and a plywood cricket rearing pen to house the crickets. Table B4.1.1 shows the costs and quantities needed for each item. In total, the project pays US$220 per household starter kit. Most of the items in the kit are long-lasting structures and materials, such as the plywood pens and plastic incubation boxes for egg hatching. Only the substrate and cotton wool are meant for single use.

TABLE B4.1.1 Items Supplied to Cricket Farmers in Kenya's Kakuma Refugee Camp

2019 US dollars

Item	Units	Price per unit (US$)	Number of units supplied to each beneficiary	Price (US$)
Plywood pen (8x2x2 feet)	Pens	59.40	2	118.80
Incubation boxes	Boxes	9.90	5	49.50
Carrier bag	Bags	0.20	5	0.99
Egg trays	Trays	0.07	80	5.54
Water sprayer	Bottles	0.01	1	0.01
Disposable plates	Plates	0.25	5	1.24
Chicken feed	Kilograms	0.69	30	20.79
Cotton wool	Rolls	3.96	6	23.76
Total cost				220.63

Source: J. Kinyuru, personal communication, 2019.

As the insect sector evolves, commercial producers will feel pressure to introduce product traceability, which is the practice of documenting how insects are fed and handled, particularly postharvest as insect products pass through the supply chain. Traceability can ensure safety and quality. This will particularly affect the types of substrates producers use to feed their insects. As insect quality and traceability improve, commercial producers will then segment the insect market by introducing differentiated products (figure 4.7). These will

TABLE 4.3 Inputs and Cost of Producing 1 Kilogram of Crickets in Kenya

2019 US dollars

Input	Initial investment required to produce 5 kg of crickets (wet weight)		Cost of producing 1 kg of crickets (wet weight)
	Amount	**Cost (US$)**	**Cost (US$)**
Incubation boxes	1	1.10	0.22
Egg trays	40	0.12	0.02
Disposable plates	6	1.00	0.20
Eggs (plates)	8	4.00	0.80
Feed (kg)	8	6.40	1.30
Water (liters)	10	1.00	0.20
Cotton wool (grams)	400	2.00	0.40
Total		15.62	3.14

Source: Interview with J. Kinyuru, 2019.
Note: Cotton wool has a lifetime of six months, the incubation boxes have a lifetime of 15 years, and the egg trays have a lifetime of 5 years. kg = kilograms.

FIGURE 4.7 Price Changes from Market Segmentation and Outsourcing Production to Small-Scale Insect Producers

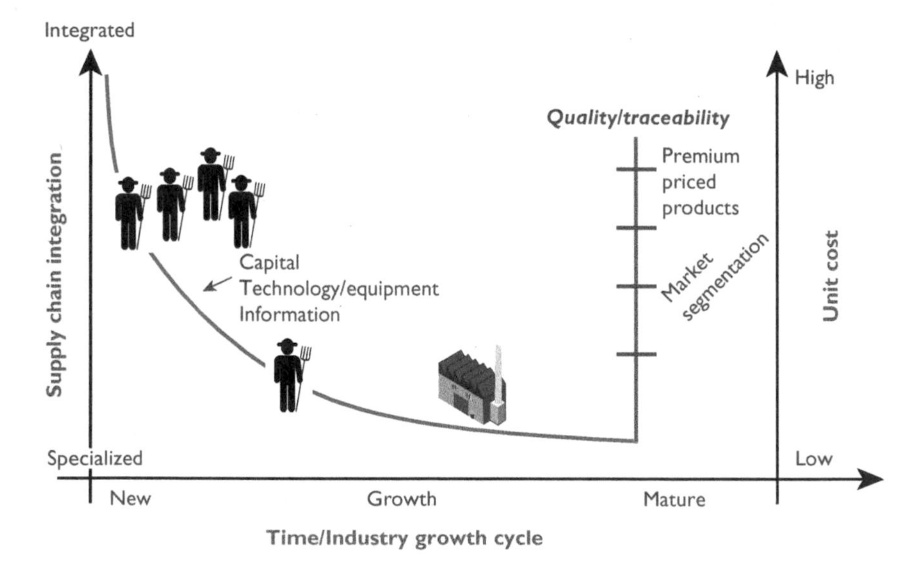

Source: Original figure for this publication.

include premium priced products, such as insects reared on substrates from certified organic crops or other high-value inputs. Eventually, commercial producers in Africa will need to comply with the same regulatory standards as those in the EU and US markets. For example, all edible insect substrates must be preconsumer, meaning no postconsumer waste can be included in the substrate. These quality standards will result in a steady rise in the unit cost of commercial production, possibly to the point at which commercial producer product prices equal rural producer prices.

Market segmentation may eventually affect insect prices and production schemes in rural areas. Africa's poor transportation infrastructure to rural communities will limit commercial producers' ability to reach rural consumers. In response, commercial producers are likely to acquire or partner with successful small-scale, rural producers and introduce capital, information, and technology to rural producers. This would bring down the unit cost of insects and insect products in rural markets. Since rural consumers are typically poorer and, therefore, price sensitive, a slight reduction in insect prices would likely displace some small-scale community insect production systems. In this scenario, an out-grower scheme may develop whereby the small-scale farmer subcontracts displaced rural farmers to rear insects (figure 4.7). The displacement of small-scale producers would not necessarily end household insect production, which is used primarily for home-based insect consumption and not for selling.

EDIBLE INSECT PRODUCTION SYSTEMS

This section describes the production systems for houseflies, crickets, mealworms, silkworm chrysalids, palm weevil larvae, and BSF.

Houseflies

Houseflies have a short and simple production cycle. The housefly (*Musca domestica*, L. (Diptera: Muscidae)) is found everywhere in the world that humans settle (van Huis et al. 2020). Adult females lay their eggs in moist, nutrient-rich environments—such as food waste and manure—and can lay up to 500 eggs in their lifetime. One must only leave substrates open and wild houseflies will naturally lay their eggs there. The 3- to 9-millimeter larvae (maggots) hatch within 8 to 20 hours and feed immediately on the substrate on which the eggs were laid. Larvae go through three instar stages over three to five days and then pupate. After two to six days, pupae develop and emerge as adults. The housefly's adult life stage lasts up to 25 days. The housefly has a short larval growth phase and a long adult phase during which it actively feeds. This contrasts with the BSF, which has a short adult phase during which it does not feed and consumes only small amounts of water. Houseflies are disease vectors, so housefly mass rearing structures must follow correct procedures to ensure that the flies are well-contained.

Housefly larvae that are reared on waste substrates show potential as an efficient animal feed source in Africa (Kenis et al. 2018; Koné et al. 2017; Sanou et al. 2018; Pomalégni et al. 2017). In Benin, a survey shows that 41 of 714 poultry farmers used housefly larvae to feed their poultry. Most of the farmers using housefly larvae were in southern Benin and, on average, were better educated than other farmers. These farmers also tended to raise larger chicken flocks in a confined system, in contrast to open scavenging systems, and also had higher poultry-related incomes than poultry farmers who did not use housefly larvae. The farmers who used housefly larvae also tended to use other innovative insect feeds for their poultry, including termites (Pomalégni et al. 2017). Houseflies' larvae can biodegrade manure, fish offal, slaughter blood, cereal or legume waste, and other low-quality organic matter streams. These streams are then rapidly assimilated into the insect's biomass. One kilogram of dry organic waste can be turned into 150 to 200 grams of fresh housefly larvae. Moreover, the larvae's short life cycle limits the amount of substrate it can consume; therefore, the same substrate may be used for larvae production for two cycles (Ganda et al. 2019).

Crickets

Crickets have become a popular insect to farm because they have a good flavor and can be domesticated. Wild crickets are consumed in many traditional diets in Asia and Africa because they are tasty and easy to prepare. People consume full-sized adult crickets, which resemble small shrimp in size and appearance, unlike other edible insect species, which people consume as larva. The cricket species that are most ideal for farming are known as colonial crickets because they live in large groups, or colonies, and can be kept in high densities. Other wild cricket species have solitary behavior and are not suitable for domestication. Crickets belong to the Orthoptera insect order. As such, crickets hatch from eggs and develop from a nymph stage to a mature adult stage when, stepwise, they molt their chitin exoskeleton (instars). The cricket production cycle has three stages: (1) hatching eggs, (2) growing hatched nymphs to maturity for harvest, and (3) mating and egg laying for the next cycle.

Cricket farming has several key characteristics. Cricket farming structures comprise a series of containers in which batches of crickets are produced to maturity. Different types of cricket farms are shown in photo 4.1. In simple systems, cricket eggs are hatched and nymphs grow to adults in the same container. In more advanced systems, the egg hatching and first instar stages of nymphs are kept in different containers to adjust the temperature and humidity to create an ideal environment for each stage. The productivity of the cricket farming system is determined by three factors: the egg hatching rate, the nymphs' survival rate, and the growth rate of nymphs to mature adults for

PHOTO 4.1 Examples of Cricket Farms

a. Good agricultural Practices–certified *Acheta domesticus* cricket farm in Thailand

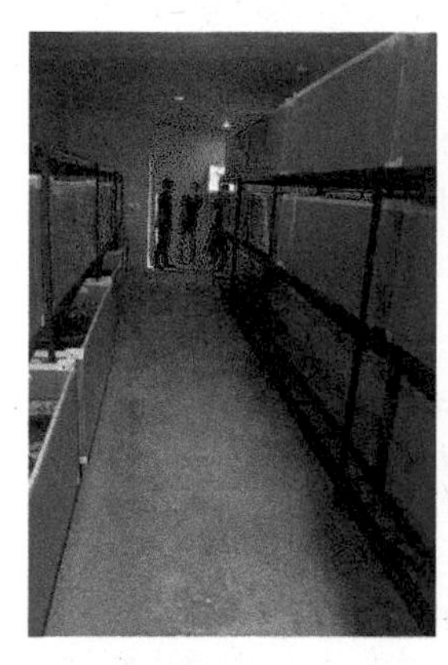

b. *Gryllus bimaculatus* cricket farm in northeast Thailand

c. InsectiPro's cricket farming system in Kenya

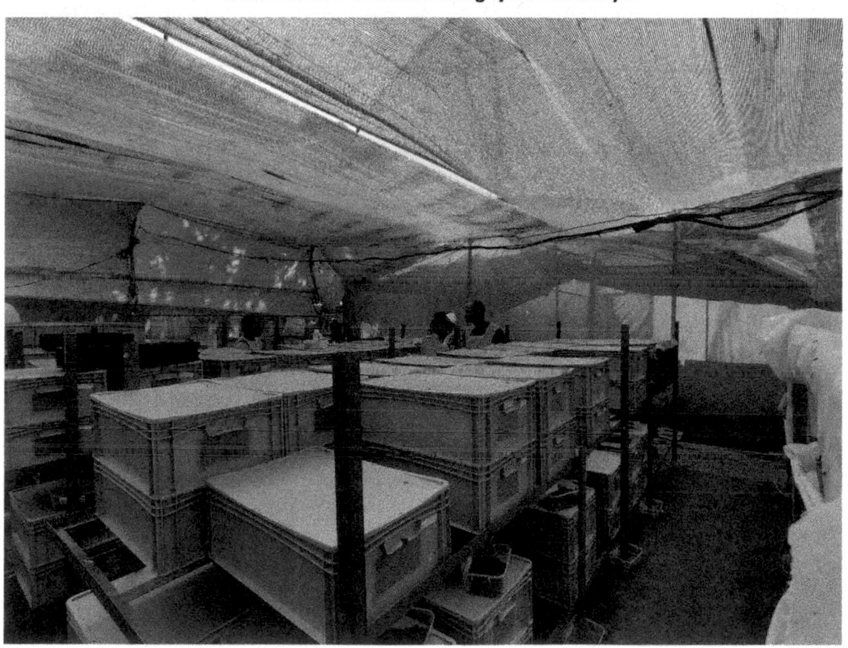

Photographs (panels a and b) © Nanna Roos / University of Copenhagen. Used with the permission of Nanna Roos. Further permission required for reuse. (panel c) © Dave de Wit / InsectiPro. Used with the permission of Dave de Wit. Further permission required for reuse.

harvest. The cricket's substrate and environmental conditions, primarily temperature and humidity, are also key factors. According to studies from Kenya (Orinda et al. 2017; Kinyuru and Kipkoech 2018) and Cambodia (Miech et al. 2016), and studies carried out under laboratory conditions (Morales-Ramos, Rojas, and Dossey 2018), the optimal temperature for hatching and growing crickets ranges between 25°C and 30°C.

Feeding crickets requires meeting the cricket's nutritional requirements from available substrates. In Thailand, cricket feed has a similar nutritional composition to common chicken feeds (Halloran 2017; Halloran et al. 2017). However, since this feed is suited for chickens, it is not fully optimized for the specific nutritional needs of crickets. Crickets can also consume fruits, vegetables, and even weeds. Public and private sector feeding experiments are identifying the nutritionally optimal feed sources for crickets (Magara et al. 2019; Dobermann, Michaelson, and Field 2019; Neville and Luckey 1962; Veenenbos and Oonincx 2017).

Cricket value chains are complex and varied. The general value chain for farming crickets is illustrated in figure 4.8. Farmers sell adult crickets for direct human consumption or further processing and can sell cricket eggs and nymphs to other farmers to start new colonies or add to existing colonies to prevent inbreeding. Farmers can also sell cricket frass, a production by-product that can be turned into a biofertilizer, to other farmers to fertilize crops or vegetable gardens. These value chains vary by country. In Uganda, farmers sell their fresh crickets, both *Acheta domesticus* and *Gryllus bimaculatus*, to a research project for a fixed price. In Madagascar, farmers dry and pulverize the crickets before selling the powder to a nongovernmental organization, which uses it to increase the protein and nutrient content in foods for undernourished children. In all the surveyed cricket farming countries, farmers sell cricket eggs to other current or potential insect farmers. In the Democratic Republic of Congo and in Kenya, farmers sell crickets to wholesalers, who in turn sell them to vendors. The vendors can then sell the insects fresh or cooked. In Kenya, farmers sell crickets to fish breeders and chicken farmers.

Different cricket management and processing practices create products of various uses and properties. In Thailand, it is a standard practice to feed pumpkin to crickets during the last days before harvest to improve their taste and golden color from the pumpkin beta-carotene. Thai cricket farmers also manage their colonies to have more females than males, which increases the market value because females carry eggs inside, making them tastier. Crickets are often dried and pulverized into flours of different properties. For example, the Gryllus species' flour is darker than the house cricket's flour. The most common cricket flour is processed as whole flour, for which the entire cricket is ground up. Cricket flour can be defatted, separating the protein from the fat, to improve the flour's quality for certain applications

FIGURE 4.8 Cricket Value Chain

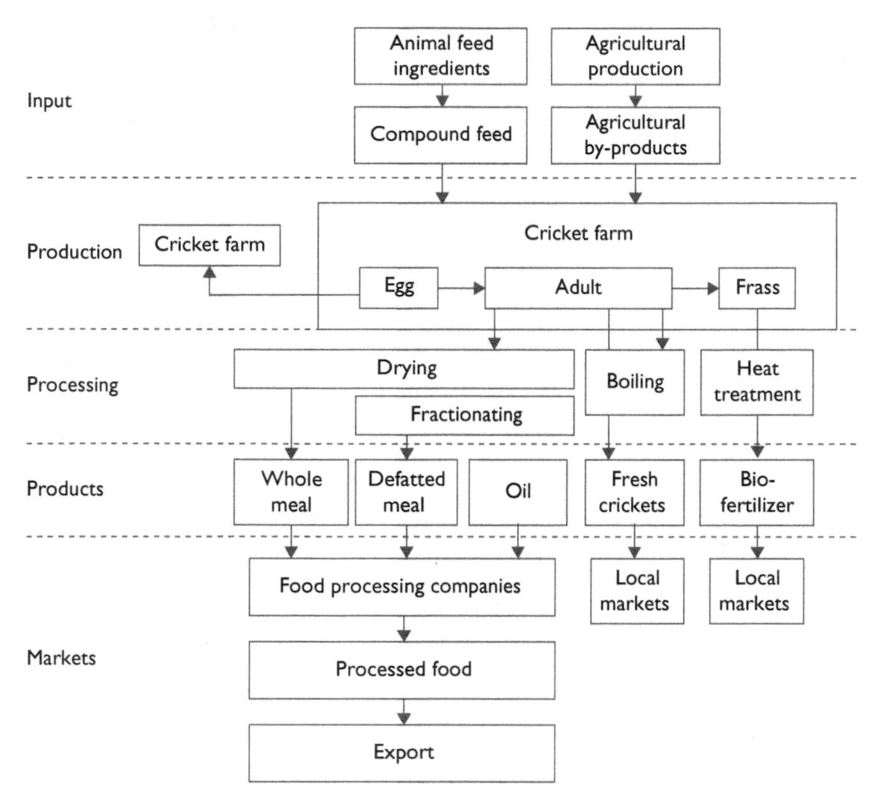

Source: Original figure for this publication, using interviews, observations, and published information.

(Sipponen et al. 2018). Cricket flour offers a high-protein-content substitute for wheat or other kinds of flours with lower protein content (Homann et al. 2017). Processing—such as drying, roasting, frying, freezing, or baking—adds value and shelf life to whole crickets. An insect farming expert in Zimbabwe described this: "As more farmers venture into cricket production, there is increasing interest from the confectionary industry for cricket powder for use in baking."

The edible cricket sector is at various stages of development in different parts of the world. Thailand has a fully developed, commercial cricket farming sector with advanced value chains. This is unique in the world. The sector produces a large variety of products, including cricket snack foods. Thailand's dynamic cricket value chain includes chefs and restaurants using crickets to develop new dishes and insect-themed meals. Since 2014, cricket farming has been introduced experimentally by universities in western Kenya (Jaramogi Oginga Odinga University of Science and Technology) and central Kenya (Jomo Kenyatta University of Agriculture and Technology) (Kinyuru and Ndung'u 2020). In Kenya and Uganda, international donor projects are developing and subsidizing semi-commercial cricket production and value chains and marketing. The supply of crickets to the markets in these countries remains scattered, and supply chains are still underdeveloped as the cricket sector grows.

Mealworms

Mealworms are easy to farm. Mealworms are the larval form of the mealworm beetle. The yellow mealworm (*Tenebrio molitor*) is the most widely farmed mealworm species in the world, although the lesser mealworm (*Alphitobius diaperinus*) is increasingly farmed in Europe. Mealworms are easy to contain. The mealworm beetle lays many eggs, it is easy to care for the larvae, and the larvae feed on many organic substrates. The mealworm beetle insect order, or *Coleoptera*, has four life cycle stages: egg, larvae, pupae, and mature adult. The evolution of the insect is considered a complete metamorphosis. The mealworm production cycle generally follows the insect's life cycle. The production stages include (1) hatching eggs, (2) growing larvae for harvest, (3) retaining larvae pupae for breeding, (4) allowing pupae to mature to adults, and (5) mating and egg laying for reproduction. Typical mealworm production systems, which are popular in Korea and European nations and to a limited extent in Africa (photo 4.2), are manually managed and include easily handled, stacked crates. Farmed mealworms are generally fed wheat bran or other cereal substrates, supplemented by vegetable substrates. Mealworms have short value chains because processing adds little value. Instead, these insects are usually transported and sold directly to markets. That said, multiple processing technologies have been experimentally applied to process fresh or dried mealworm into minced meat-like products or other products (Stoops et al. 2017; Tonneijck-Srpová et al. 2019). Drying is the most common type of mealworm processing and can be done in a conventional or microwave oven. Microwave drying preserves the mealworm's color but can make the insect greasy and unpleasant to consume.

Mealworms' potential as food and animal feed is nascent. In Africa, mealworm farming produces human food and animal feed, although at a very limited scale. Historically, people have bred mealworms for pet food for captive fish, birds, and reptiles. Pet food is not under food safety regulations, and the scale and nature of mealworm production have not been recorded historically. Mealworm production has only recently moved into the market of human foods and animal feeds, initially using mealworm production systems already in place for the pet food market. In Europe, investments in research

PHOTO 4.2 Examples of Mealworm Farms

a. Mealworm (*Tenebrio molitor*) production stacked crates in the Republic of Korea

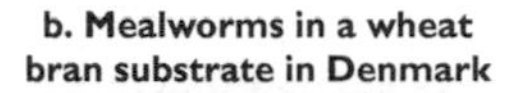

b. Mealworms in a wheat bran substrate in Denmark

c. Mealworm crate system in Denmark

Photographs (panel a) © Nanna Roos / University of Copenhagen. Used with the permission of Nanna Roos. Further permission required for reuse. (panels b and c) © Jonas Lembcke Andersen / Danish Technological Institute. Used with the permission of Jonas Lembcke Andersen. Further permission required for reuse.

and development of the yellow mealworm (*Tenebrio molitor*) and the lesser mealworm (*Alphitobius diaperinus*) have helped upscale mealworm production to modern automated systems, in which fresh mealworms are sold for US$1.10/kg (Macombe et al. 2019). Mealworm farming has received less attention in Asia than in Europe and North America. Korea is the exception because of the government's promotion of insect farming (MAFRA 2019). As a result, the government of Korea reported that there were more than 200 mealworm farms in the country in 2019. By contrast, mealworm farming in Africa is limited compared with that of other insects like crickets, BSFL, or palm weevil larvae.

Silkworm Chrysalids

The silk moth (*Bombyx mori*) is the most ancient domesticated insect, and silk production is carried out in an industrialized sector. Silk moths produce the most valuable silk but are dependent on a single feed source, mulberry leaves, which are usually planted on the edges of vegetable fields. By contrast, the eri moth (*Samia ricini*) produces a less valuable silk but can feed on various leafy substrates, making it easier to farm. Both moths are farmed in Africa, particularly in East and Southern Africa, although silkworm farming (sericulture) is most common in Asia.

The silk production cycle mirrors the silkworm's life cycle. Silkworms go through four stages of development: egg, larva, pupa, and adult. Usually, there are three days of dormancy between each stage. The larva stage is the silkworm caterpillar, and the adult stage is the silkworm moth. During the egg stage, the female butterfly can mate with a male and lay 400–500 eggs at a time, called a cell. It takes about three to five days for the eggs hatch. During the larva stage, silkworms grow rapidly, so they must molt (shed their skin) four times while growing. The design of the silk farms can vary depending on local traditions and the availability of local materials. For example, in Madagascar, newly hatched juveniles are transferred to racks made with bamboo or round wood and placed on top of a layer of thatch or paper. The larvae are fed three to four times a day. On the 31st day, each larva begins to weave a cocoon. The farmer then places dry straw or dry fern below the larva so it can settle and weave the cocoon for seven to eight days. During this time, the chrysalid, or pupa, forms inside the cocoon. If the chrysalid is left alive to evolve for 15 days, it excretes an enzyme that allows it to escape the cocoon as a moth. If the farmer wishes to produce silk, the farmer kills the pupae before they can produce the enzyme. The farmer then boils the cocoons and extracts the silk. The average Malagasy farmer produces approximately 80 kg of cocoons per year. Once all the silk is extracted, the farmer removes the chrysalids from the boiled water. The dead chrysalid, or pupa, is a by-product of silk production, which is fried and sold as a nutritious snack. A few restaurants in Madagascar use chrysalids in their dishes. In total, silk production takes about 37 days from egg hatching to silk harvesting. In Madagascar, farmers sell silk to artisan spinners directly or through middlepersons.

Silkworm farming has a complex value chain. In Madagascar, the research team observed that silkworm producer associations or cooperatives sell silkworm cocoons—from which the silk is spun—directly to artisan spinners or weavers. By contrast, individual silkworm producers must first sell their cocoons to middlepersons, who then sell the cocoons to artisan spinners. The spinners extract the edible silkworm chrysalises and sell them to other middlepersons, who then deliver the chrysalises to retailers in popular markets in Antananarivo and Antsirabe. Restaurants rarely buy directly from breeders.

Palm Weevil Larvae

Palm weevil farming replicates wild harvesting systems but with fewer environmental impacts. Palm weevil production occurs in many countries throughout West and Central Africa. Traditionally, palm weevils are collected from palm trees in the wild. This practice is destructive and can negatively affect an ecosystem. Palm weevil farming offers an alternative production system. In modern palm weevil farming systems, farmers collect adult palm weevils from the wild by trapping them in buckets laced with a pheromone (Rynchophol) and fermented fruit. The traps are inspected three times a week to collect adult weevils. These buckets are located near or inside the farmers' homes. The farmers place five pairs of adult weevils of both sexes in a bucket with about 5 kg of oil palm or raffia tree yolk, or inner core, as a feed substrate for the insects. The yolk is mechanically shredded or chopped into small pieces and then soaked in water basins. After three days, the yolk is removed and allowed to stand for two hours to drain excess water prior to using it for adult inoculation. The farmers add about 100 grams of sugar for the weevils to feed on. The weevils then mate and, after two days, lay eggs. The eggs hatch after a day or two depending on the environmental conditions. The larvae consume the feed and can be harvested after four to five weeks once they mature to the seventh or eighth instar stage or grow to about 6 to 10 grams. A portion of the larvae can pupate and grow into adulthood, thereby restarting the production cycle. Studies in Cameroon show that weevil farming requires less time and fewer resources than traditional collection and semi-farming methods. Farming systems also increase the farmers' productivity and, consequently, the total volume of the insects they produce (table 4.4).

Black Soldier Flies

BSF production systems are the fastest-growing subsector of the insect industry. BSF has become the preferred insect species for large-scale insect protein production because of its flexible feeding strategy, growth efficiency, disease resilience, waste management properties, and environmental sustainability. Because of these benefits, BSFL are the "black gold" of alternative feed proteins. The BSF, *Hermetia illucens*, L. (Diptera: Stratiomyidae) originates from the Americas but is now common worldwide (Martínez-Sánchez et al. 2011; Wang and Shelomi 2017). BSF belongs to the Diptera, or true fly, order. BSF has a short adult life span of only a few weeks during which the fly does not feed. Adults form mating swarms and deposit a single clutch of eggs two days or so later. BSF eggs hatch after approximately four days and the resulting larvae (BSFL) require 10 to 14 days to reach the prepupal stage, which is the optimal time to harvest. The resulting pupae require two weeks to reach the adult stage. BSFL typically weigh 120–180 milligrams at the time of harvest. This means that it takes approximately 5,500 to 8,300 larvae to equal 1 kg.

TABLE 4.4 Productivity of Different African Palm Weevil Farming Systems

Production system	Quantity of raffia used	Productivity
Traditional gathering	1 raffia stem of 2 to 3 meters	35 larvae
Semi-farming	1 raffia stem of 2 to 4 meters	50 larvae
Farming	Less than ¼ stem	69 larvae

Source: Muafor et al. 2015.

BSFL convert a significant amount of substrate and are not considered pests. The BSFL feed on most organic substrates, including fruits, vegetables, cereals, tubers, and legumes. But they can also feed on organic wastes, such as animal manure (Li et al. 2011; van Huis 2019; Sheppard 1983) and fecal sludge (Lalander et al. 2013). The potential of using BSFL to break down organic waste has been explored for decades (Sheppard 1983). Only the BSFL feed; the adult BSF does not have a mouth, and so it does not feed or bite. As such, the BSF is not considered a pest.

BSFL production is also used for waste management in many parts of the world. BSFL can convert organic waste into different end products, such as oil, protein, and biofertilizer. In Australia, Korea, and Malaysia, industrial-scale BSFL are used to reduce urban wastes. The success of these operations depends on the safe and efficient handling of large volumes of waste. In China, Kenya, and South Africa, BSFL are used to reduce manure from large-scale livestock production and fecal sludge from cities. In Europe, the main concern with producing BSFL feed from manure is making sure the feed meets safety requirements (EFSA 2015), while in Asia, the concern is converting large amounts of organic wastes from crowded urban settlements (Wang and Shelomi 2017). Consequently, facilities in the European Union and North America are designed to have BSFL digest preconsumer food waste and market the resulting larvae for poultry or aquaculture feed, while facilities in Asia are designed to convert large volumes of mixed wastes into insect biomass. For example, over a 15-year period, China developed industrial-scale BSFL plants with the capacity to convert more than 100 tons of organic waste per day. Table 4.5 shows the estimated values from BSFL converting fecal sludge into protein, biofuel, and biofertilizer in three African cities (Diener et al. 2014).

BSFL grow quickly, building up fat, protein, and chitin. The BSFL's growth period is only two weeks from hatching to harvest. During this two-week period, a larva weighing less than 1 milligram grows to 180 milligrams or more. This growth rate is almost two times faster than the average growth rate of a broiler chicken. This would correspond to a chicken weighing 40 grams when it hatches and growing to more than 700 grams in two weeks. The BSFL's growth varies with the substrate it is fed and the abiotic environment

TABLE 4.5 Value of BSFL Converting Fecal Sludge into Different End Products in Three African Cities

End product	Dakar, Senegal	Accra, Ghana	Kampala, Uganda
Dry fecal sludge legally discharged per day (tons, thousands)	6	26	16
Value of BSFL converting this sludge to protein for animal feed (US$, thousands)	40,000	235,000–255,000	129,000
Value of BSFL converting sludge to fuel-biogas and selling the remaining residue as biofertilizer (US$, thousands)	—	248,000–258,000	159,000
Value of BSFL converting sludge to biofertilizer (US$, thousands)	12,000	54,000–134,000	81,000

Source: Diener et al. 2014.
Note: BSFL = black soldier fly larvae; — = not available.

(temperature and humidity) in which it is grown. BSFL are high in protein and chitin. The fat and protein contents of harvested BSFL also depend on the substrate and, possibly, the time of harvest and different genetic traits (Wang and Shelomi 2017), but this has not been standardized across BSLF feeding studies. Wang and Shelomi (2017) found that BSFL, on average, produced 37 to 45 percent (40.8 ± 3.8 percent) protein (dry weight) and 20 to 36 percent (28.6 ± 8.6 percent) fat (dry weight) across 22 studies. This relatively high fat content is a general biological characteristic of insect larvae. The protein content is on par with other animal feeds like fishmeal (Shumo et al. 2019). BSFL, like all insects, also contain chitin, which originates from the insect's exoskeleton. Chitin is functionally regarded as a dietary fiber, although studies suggest that it may also have probiotic properties benefiting animals' gut health (Selenius et al. 2018).

BSFL production can be carried out in fully or semi-managed systems. In general, BSFL production includes three phases: (1) egg production, (2) egg hatching and larva growth, and (3) harvesting. BSF breeding and egg laying—called oviposition—require a netted cage in which the adult flies swarm and mate. After mating, the eggs are mixed with an organic substrate and hatch after four days. Next, the larvae grow for about 12 to 14 days. Typically, this process takes place in a stacked crate system (see photo 4.3 for examples of BSF production systems). Last, the BSFL farmers harvest the larvae by removing them from the unconsumed substrate. Alternatively, BSFL can be reared in simpler, semi-managed systems with open containers of organic substrate that attract wild BSF (Nyakeri et al. 2016). However, in unmanaged systems, the BSFL will "self-harvest," or naturally separate from the substrate at the prepupal stage. This characteristic skips the step of separating the larvae from the substrate;

PHOTO 4.3 Examples of BSF Production Systems

a. Automated system in Denmark

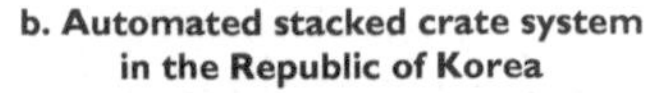

b. Automated stacked crate system in the Republic of Korea

c. Net-breeding system in Kenya

d. Small-scale system in Kenya

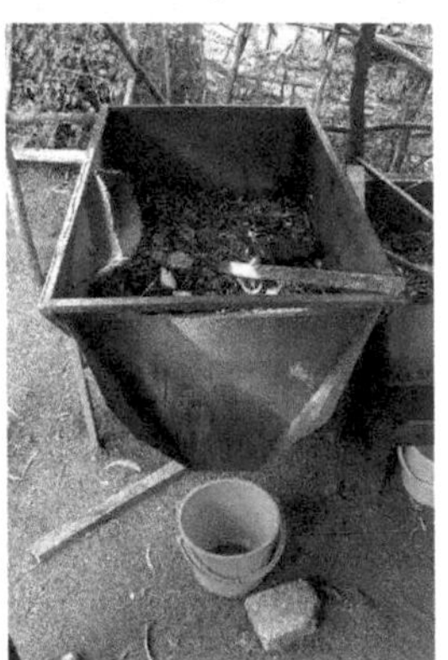

however, leaving the larvae to grow unmanaged to the prepupae stage is not optimal because only some larvae will leave the substrate. Furthermore, the prepupae will have lost mass and have a more sclerotized cuticle. As a result, fully managed systems are more efficient for producing BSFL.

Breeding BSF is difficult, but new methods are making it easier. Managing the BSF's full life cycle is challenging because it is hard to control the adult flies (Sheppard et al. 2002). Flies will mate only in daylight, and oviposition requires a suitable material to attract the flies during a narrow time frame in their short life span. Researchers and commercial producers have developed various models to overcome these challenges, including using artificial light to replace daylight, identifying alternative substrates to attract flies to stimulate oviposition, and designing containment structures that are optimal for egg laying and hatching rates.

BSFL are being commercialized at an industrial scale. The advancements in controlling BSF breeding have made large-scale production systems possible. Globally, several companies have invested in large-scale BSFL production. New technologies for industrial-scale BSFL production are emerging quickly. The World Intellectual Property Organization's portal for patents shows that there are 259 entries for "black soldier fly."[4] Industrial-scale production plants with semi- or fully automated production can handle up to 1,000 tons of organic waste per day. The first large-scale industrial production system for BSFL was established in 2015 (Drew and Pieterse 2015). However, production capacity is largely unknown because companies have not made this information public nor do any independent organizations seek to compile these data. That said, some market analysis companies have attempted to map the BSFL sector and rank the leading companies by the estimated production capacity (Persistence Market Research 2019). This research shows that several producers have invested in industrial-scale BSFL production facilities in recent years, including Protifarm (the Netherlands), AgriProtein (South Africa/United Kingdom), Ynsect (France), Protix (the Netherlands), C.I.E.F. (Korea), and Enterra (Canada) (see table 3.7). Other companies have established the capital investments needed to reach industrial production capacity, including Enviroflight (United States), Nutrition Technologies (Malaysia), Goterra (Australia), and ENORM (Denmark).

BSFL's main market is animal feed. BSFL can be processed into meals that are fed to animals. In Europe, Canada, and the United States, initial efforts to market BSFL were made in aquaculture, or as a feed ingredient for selected fish species. More recent efforts have expanded the market to pet foods in the European Union and poultry in Canada and the United States. Pet and poultry markets support higher price points than the aquaculture market, making BSFL-based meals economically viable for BSFL producers. Moreover, fishmeal prices are rising, and if this continues, the viability of BSFL meal replacing fishmeal will become more realistic. A female farmer in Kenya commented, "Feed makes up 60 to 70 percent of the cost of animal production and the cost of protein makes up 70 percent of the cost of feed. So, we started looking at alternative options for protein and landed on the black soldier fly. We started in November 2018 with 2 kilograms of BSF and are now [November 2019] at 3 tons of daily production." In Africa, BSFL substitutes for animal feed ingredients, like fishmeal, are more viable than in Europe. In Kenya, BSFL meal is competitive with both fishmeal and soybean meal on a protein per unit price basis (table 4.6). And since BSFL production systems in Africa are less automated than those in Europe and North America, start-up and production costs can be lower, translating into higher profit margins initially.

BSFL processing can separate elements of the insect for different purposes (figure 4.9). Maintaining the oil fat in BSFL increases BSFL meal's rancidity

TABLE 4.6 Prices of Protein Sources for Animal Feed Available in Kenya

2018 US dollars

Product	Protein content (%)	Price[a] (US$/kilogram)	Protein unit price (US$/kilogram)
Black soldier fly larvae meal	66	0.85	1.29
Fishmeal	51	1.00	1.96
Soybean meal	49	0.75	1.53
Cottonseed meal	44	0.28	0.64
Sunflower meal	32	0.23	0.72

Source: Vernooij and Veldkamp 2019.

a. Values are per kilogram of product standardized to 10 percent moisture and converted to US$ from original values in Kenyan shillings.

during storage. Therefore, pressing, boiling, or organic solvents are used to separate the BSFL oil, or fat, from the meal. This preserves the meal's shelf life. The processing technology used for defatting depends on the available equipment and the desired quality of the end product. The processing technology affects the BSFL's protein content and nutritional value and the dry protein residue's bioactivity. The composition of frass—or BSFL manure—depends on the substrate's chemical composition and the larva's ingestion and retention of nitrogen for protein synthesis. BSFL frass can serve as a valuable biofertilizer that improves soil health by adding carbon and nutrients to depleted soils. There are various processing methods for BSFL, including for wet or dry BSFL. Processing wet larvae allows the feed processer to separate the chitin as well, enhancing the protein content of the defatted product. Processing dry larvae is technologically a simpler and cheaper approach when separating protein from fat. The chitin remains in the protein fraction during the separation.

MODELING THE POTENTIAL OF BSF IN ZIMBABWE

This case study modeled the BSF supply chain to approximate BSFL production levels in Zimbabwe given different crop substrates. Five key crops grown in Zimbabwe—maize, wheat, groundnut, soybean, and sugarcane—were selected for the model. The model uses the five-year average, from 2013 to 2017, of each crop's national production totals and annual harvested area, as reported by the Food and Agriculture Organization of the United Nations (FAO), to determine the average yield per crop.[5] Based on these averages, the model calculates crop-associated waste and losses[6]—namely, residue, postharvest loss,

FIGURE 4.9 BSF Value Chain

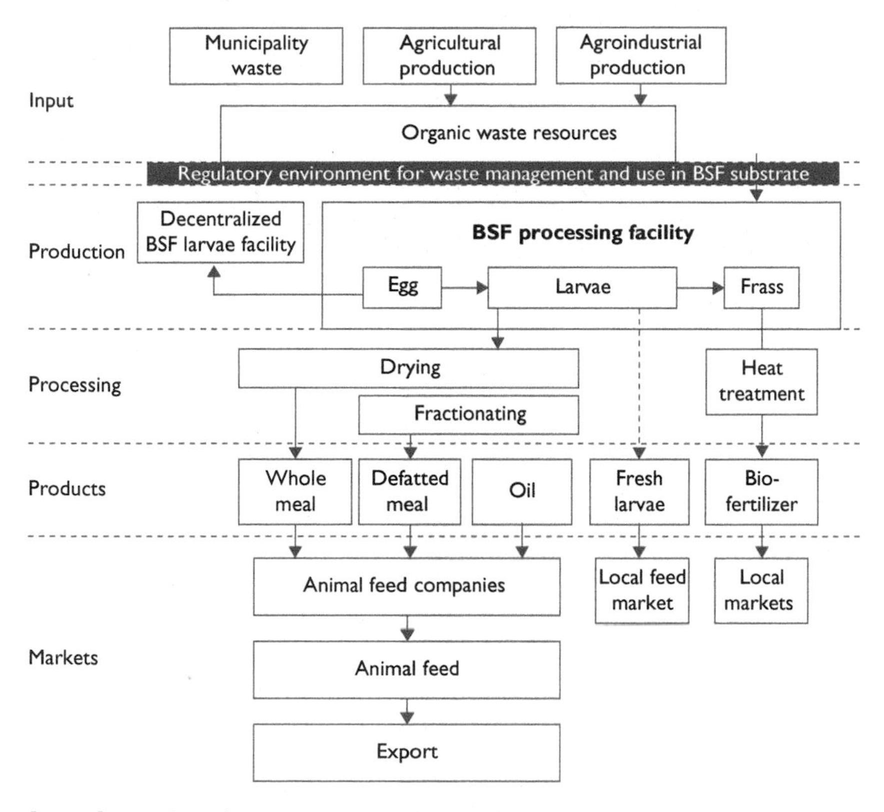

Source: Original figure for this publication, using interviews, observation, and published sources: Sheppard et al. 2002; Wang and Shelomi 2017.
Note: BSF = black soldier fly.

processing waste, distribution loss, and consumption waste—along with each crop's respective food supply chain and waste stream. Using the crop waste as BSFL substrate, various conversion factors are employed to calculate theoretical quantities of BSFL frass and dry meal that could be produced on a per-crop basis. These conversions include (1) residue to BSFL, (2) process waste to BSFL, (3) residue to frass, (4) amino acid to frass, and (5) BSFL to dry meal. However, regardless of the crop type, and consistent with generally accepted guidelines,[7] the model uses a low range of 10 percent and a high range of 30 percent for the aforementioned conversions with the exception of BSFL to dry meal, which is assumed to be 34 percent for all crops. It was necessary to approximate these ranges because no studies have determined BSFL's crop-specific rates of converting crop wastes to frass and dry meal. The model also estimates the crude protein content of dry meal to be 40 percent regardless of the crop. These conversion factors are listed in table 4.7. The model then compares the total crude protein derived from the dry meal[8] against the protein requirements of certain livestock: pigs, goats, chickens, and fish from aquaculture.

The model was extended beyond Zimbabwe to analyze Africa as a whole and 10 additional African countries individually. The analysis for all of Africa included the same five crops studied in Zimbabwe—maize, wheat, soybean, groundnut, and sugarcane—to determine the potential production volume of BSFL and frass and the impact of each. The assessment of individual African countries included the 10 countries with the largest agricultural economies on the continent. These are, from largest to 10th largest, Nigeria, Kenya, the Arab Republic of Egypt, Ethiopia, Algeria, Tanzania, Morocco, Sudan, Ghana, and Angola. The analysis added cassava to the list of crops to reflect the agriculture sector in the 10 countries, although the analysis for each country was limited to only five of the six crops—the five with the highest production tonnage for each country (annex 4A provides details of the analysis and the full results).

Achieving maximum conversion rates depends on how substrates are prepared. Crop residues, for example, may be fibrous and may not be fully consumed by BSFL. Therefore, grinding and fermenting fibrous materials before feeding them to BSFL allows the BSFL to consume them more completely, thus raising the substrate-to-BSFL conversion. Considering that BSF breeding is a nascent industry, more studies are required to establish best practices for crop waste preparation, determine crop-specific conversion rate estimates, and estimate more accurately BSFL's final frass and crude protein outputs.

Other wastes can be used as BSFL substrates but are not included in the model. These substrates include crops other than the five used in the modeling, including various types of processing waste. For example, rotten vegetables are ideal for BSFL consumption and can be collected from vegetable processing plants. Brewer spent grains (BSG) are another potential substrate for BSFL. BSG are an excellent source of crude protein for livestock and are in high demand for that reason. It is more efficient to feed BSG directly to livestock than to convert them to protein through BSFL. However, the shelf life of BSG is only a day or two at ambient temperatures or two or three days if kept refrigerated. Spoiled BSG should not be fed to livestock, but the

TABLE 4.7 BSF-Related Conversion Factors

BSFL to dry meal	Meal crude protein content	All other conversion factors[a]	
		Low	High
34%	40%	10%	30%

Source: Original table for this publication, using J. K. Tomberlin, personal communication.
Note: BSF = black soldier fly; BSFL = black soldier fly larvae.
a. Conversions of all crop-related wastes to BSFL and all crop-related wastes to frass.

bioremediation ability of BSFL makes spoiled BSG an ideal BSFL substrate. In short, fresh BSG are ideal for livestock, but spoiled BSG are ideal for BSFL. BSFL can also feed on animal and human waste, as discussed in the previous section, but manure and human fecal waste are not considered substrates in the model because they can potentially introduce heavy metals, salmonella, or other pathogens.

These modeled projections are not realistic in the short term for several reasons. First, initially, few farmers are likely to participate in BSF breeding. As a result, BSFL production will be limited until the sector gains wider traction. Second, it will take time to educate farmers on the benefits and techniques of BSF breeding. Farmers can learn the industry from other farmers or from technical agricultural extensionists, but this will require time. Third, BSF farms will require labor, which may be hard to attain before farms reach scale and can pay consistent wages. Fourth, crop wastes can be used for other purposes in addition to BSF breeding, so even if 100 percent of crop wastes were recovered, it would be unlikely that all of it would be used as BSFL substrate. To determine whether a particular crop residue serves better as BSFL substrate or something else would require a cost-benefit analysis.

Maize

Figure 4.10 shows the food supply chain and associated waste stream for maize in Zimbabwe. The following analysis determines BSFL meal and frass production based on the five-year (2013–17) averages for total annual maize production (735,560 tons[9]) and harvested area (1,133,690 hectares (ha)). The numbered bullets below show the kilogram per hectare calculations for each type of maize-associated waste. These are then multiplied by the total area under cultivation (hectares) and divided by 1,000 (kg/ton) to determine the total tons per hectare per waste type. Table 4.8 summarizes the total maize-derived wastes recoverable and suitable as substrates for BSF breeding. Table 4.9 applies the 10 and 30 percent conversion rates to the wastes in table 4.8 to calculate the rate of converting maize wastes to BSFL meal and frass for biofertilizer.

1. Based on the five-year average of maize production and area planted, the average maize yield is 651 kg/ha and the associated crop residue, or stover, amounts to 579 kg/ha. Stover and some other crop residues are used for various purposes. For example, a study of 310 western Kenyan farms showed that 47 percent of crop residue remains on the fields as an organic soil amendment, 25 percent is fed to livestock, 22 percent is used as cooking fuel, and the remaining 6 percent is used for miscellaneous purposes (Berazneva 2013).[10] Residues are also often burned in the field in Sub-Saharan Africa. The residues can be a clean, abundant, readily available, and essentially no-cost substrate for BSFL.

FIGURE 4.10 Maize Food Supply Chain and Annual Waste Stream, Zimbabwe

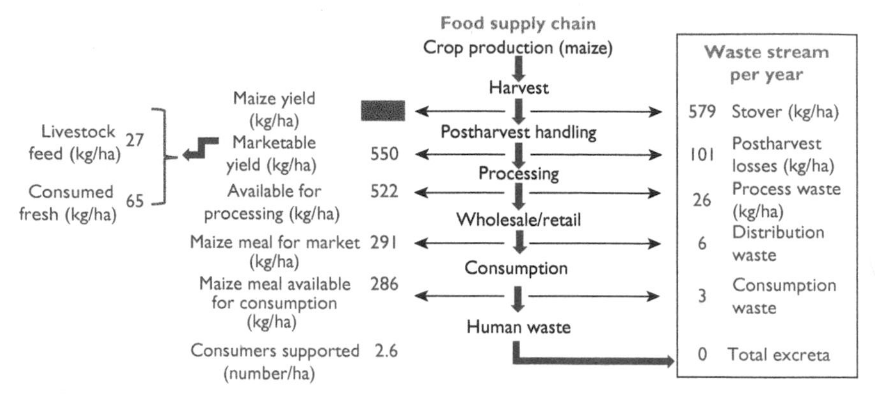

Source: Original figure for this publication.
Note: kg/ha = kilograms per hectare.

TABLE 4.8 Recoverable Maize-Derived Outputs for Black Soldier Fly Substrates, Zimbabwe

Waste/loss component	Recoverable output (kg/ha)	Total waste output (tons)
Stover	579	656,776
Postharvest losses	101	114,751
Process waste	26	29,602
Consumption waste	3	3,238

Source: Original table for this publication.
Note: kg/ha = kilograms per hectare.

2. Postharvest losses will reduce the 651 kg/ha of maize harvested by approximately 15.5 percent, or 101 kg/ha.[11] Postharvest losses provide another readily available, no-cost substrate for the larvae. However, postharvest losses may result from poor storage. If postharvest losses used as BSFL substrate are rotten or putrid, it will not harm the larvae but will likely produce a foul odor and attract houseflies, particularly in the early stages of BSF breeding.
3. The resultant maize available for the market is 550 kg/ha, of which 27 kg/ha (5 percent) are used as animal feed and 65 kg/ha are consumed fresh by humans (IITA 2020; Food Price Monitoring Committee 2003).
4. The remaining 522 kg/ha of maize are destined for processing. Processing will yield 26 kg/ha of waste, or 5 percent, which is also a source of BSFL substrate.

TABLE 4.9 BSFL, Meal, and Frass Production from Maize Output Substrates, Zimbabwe

Waste/loss component	Total waste/loss (tons)	BSFL (tons)		BSFL meal (tons)		Frass (tons)	
		High	Low	High	Low	High	Low
Stover	656,776	197,033	65,678	66,991	22,330	197,033	65,678
Postharvest losses	114,751	34,425	11,475	11,705	3,902	34,425	11,475
Process waste	29,602	8,881	2,960	3,019	1,006	8,881	2,960
Consumption waste	3,238	971	324	330	110	971	324
TOTAL	804,367	241,310	80,437	82,045	27,348	241,310	80,437

Source: Original table for this publication.
Note: BSFL = black soldier fly larvae; high = 30 percent conversion; low = 10 percent conversion.

5. Additional waste quantities, such as distribution wastes and consumption wastes, are less abundant in the supply chain, more difficult to recover, or not quantifiable. Distribution waste includes losses at wholesale markets, supermarkets, retailers, and wet markets. These wastes amount to 2 percent of the available processed maize meal (FAO 2011) and are not recovered; hence, they are not included in the maize analysis. Consumption waste includes losses and wastes at the household level and amount to only 1 percent for pulses, oil seeds, and cereals including maize meal in Sub-Saharan Africa (FAO 2011). In this case, consumption waste for maize meal is less than 3 kg/ha. However, consumption waste was included in the maize analysis because it is quantifiable and recoverable, although its amount is not significant compared with other waste sources.

Sugarcane

Figure 4.11 shows the food supply chain and associated waste stream for sugarcane in Zimbabwe. The following analysis determines BSFL meal and frass production based on the five-year averages, from 2013 to 2017, for total annual sugarcane production (3,619,823 tons) and harvested area (43,890 ha). The numbered bullets below show the kilogram per hectare calculations for each type of sugarcane-associated waste. These are then multiplied by the total area under cultivation (hectares) and divided by 1,000 (kg/ton) to determine the total tons per hectare per waste type. Table 4.10 summarizes the total sugarcane-derived wastes recoverable and suitable as substrates for BSF breeding. Table 4.11 applies the 10 and 30 percent conversion rates to the wastes in table 4.10 to calculate the rate of converting sugarcane wastes to BSFL meal and frass.

FIGURE 4.11 Sugarcane Food Supply Chain and Annual Waste Stream, Zimbabwe

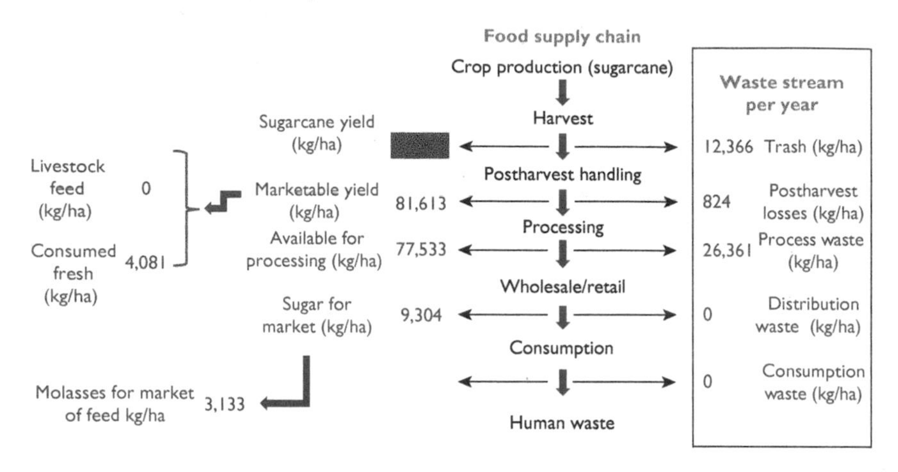

Source: Original figure for this publication.
Note: kg/ha = kilograms per hectare.

TABLE 4.10 Recoverable Sugarcane-Derived Outputs for Black Soldier Fly Substrates, Zimbabwe

Waste/loss component	Recoverable output (kg/ha)	Total waste output (tons)
Trash	12,336	542,733
Postharvest losses	824	36,182
Process waste	26,361	1,156,998

Source: Original table for this publication.
Note: kg/ha = kilograms per hectare.

1. Based on Zimbabwe's five-year average of sugarcane production and area planted, the average sugarcane yield is 82,438 kg/ha and crop residue is 12,366 kg/ha (15 percent of harvest) (Zafar 2020). This residue, referred to as trash, is often burned in the field, which is a practice considered environmentally harmful, especially considering that sugarcane is the most widely produced crop globally, at 1.9 billion tons annually (Statista 2020). As with maize stover, sugarcane trash can be a clean, abundant, readily available, and essentially no-cost substrate for BSFL.
2. Postharvest sugarcane losses are low, at 1 percent, and result in 824 kg/ha of sugarcane that can be used as BSFL substrate.
3. An estimated 5 percent of the remaining 81,613 kg/ha of sugarcane is consumed fresh (4,081 kg/ha), leaving approximately 77,533 kg/ha of sugarcane for processing.

TABLE 4.11 BSFL, Meal, and Frass Production from Sugarcane Output Substrates, Zimbabwe

Waste/loss component	Total waste/loss (tons)	BSFL (tons)		BSFL meal (tons)		Frass (tons)	
		High	Low	High	Low	High	Low
Trash	542,733	162,820	54,273	55,359	18,453	162,820	54,273
Postharvest losses	36,182	10,855	3,618	3,691	1,230	10,855	3,618
Process waste	1,156,998	347,099	115,700	118,014	39,338	347,099	115,700
TOTAL	1,735,913	520,774	173,591	177,063	59,021	520,774	173,591

Source: Original table for this publication.
Note: BSFL = black soldier fly larvae; high = 30 percent conversion; low = 10 percent conversion.

4. Bagasse and filter mud are the key components of sugarcane processing waste. Filter mud amounts to 1 to 9 percent of output, although this model uses a conservative 2 percent. Bagasse, the fibrous remains of crushed and juice-extracted sugarcane, amounts to 30 to 34 percent of output and the model uses 32 percent (Nikodinovic-Runic et al. 2013). Bagasse is used as fuel to generate steam for processing sugarcane into sugar. According to the FAO, 70 percent of bagasse can be used to generate steam, leaving 30 percent for substrate. However, this model assumes that all bagasse is substrate, although a cost-benefit comparison would be necessary to determine whether the best use of bagasse is as a fuel or substrate.
5. Additional sugarcane losses during distribution and consumption are not considered for this model.

Soybean

Figure 4.12 shows the food supply chain and associated waste stream for soybean in Zimbabwe. The following analysis determines BSFL meal and frass production based on the five-year (2013–17) averages for total annual soybean production (52,504 tons) and harvested area (43,713 ha). The numbered bullets below show the kilogram per hectare calculations for each type of soybean-associated waste. These are then multiplied by the total area under cultivation (hectares) and divided by 1,000 (kg/ton) to determine the total tons per hectare per waste. Table 4.12 summarizes the total soybean-derived wastes recoverable and suitable as substrates for BSF breeding. Table 4.13 applies the 10 and 30 percent conversion rates to the wastes in table 4.12 to calculate the rate of converting soybean wastes to BSFL meal and frass.

FIGURE 4.12 Soybean Food Supply Chain and Annual Waste Stream, Zimbabwe

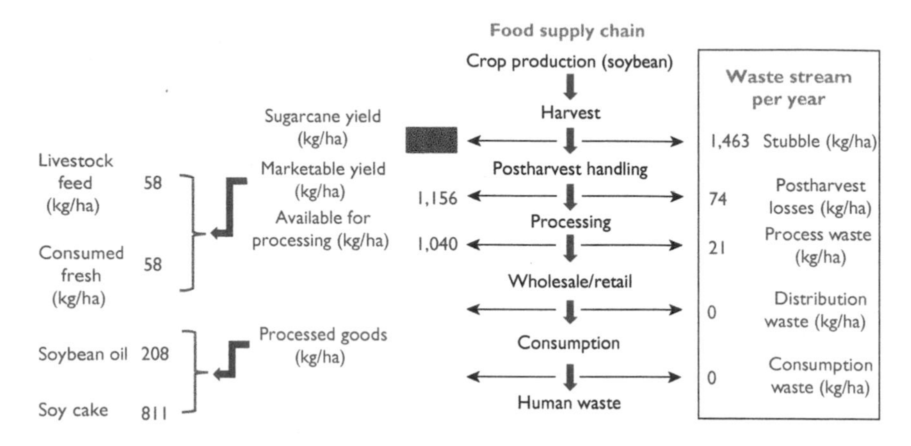

Source: Original figure for this publication.
Note: kg/ha = kilograms per hectare.

TABLE 4.12 Recoverable Soybean-Derived Outputs for Black Soldier Fly Substrates, Zimbabwe

Waste/loss component	Recoverable output (kg/ha)	Total waste output (tons)
Stubble	1,463	63,947
Postharvest losses	74	3,224
Process waste	21	909

Source: Original table for this publication.
Note: kg/ha = kilograms per hectare.

TABLE 4.13 BSFL, Meal, and Frass Production from Soybean Output Substrates, Zimbabwe

Waste/loss component	Total waste/ loss (tons)	BSFL (tons)		BSFL meal (tons)		Frass (tons)	
		High	Low	High	Low	High	Low
Stubble	63,947	19,184	6,395	6,523	2,174	19,184	6,395
Postharvest losses	3,224	967	322	329	110	967	322
Process waste	909	273	91	93	31	273	91
TOTAL	68,080	20,424	6,808	6,944	2,315	20,424	6,808

Source: Original table for this publication.
Note: BSFL = black soldier fly larvae; high = 30 percent conversion; low = 10 percent conversion.

1. Based on Zimbabwe's five-year average of soybean production and area cultivated, soybean yield is 1,229 kg/ha and crop residue, or stubble, amounts to 1,463 kg/ha, which equals 119 percent of harvest (Rees et al. 2018). Some farmers use stubble for livestock bedding since it is not as palatable as other crop residues. Other livestock farmers shred and mix the stubble with distillers' grains for livestock feed or, if stubble is available at a lower price than maize residue, use it as roughage (Rees et al. 2018). As with other crop residues, soybean stubble is a clean, abundant, readily available, and essentially no-cost substrate for BSFL.
2. Postharvest soybean losses are 6 percent, or approximately 74 kg/ha that can be used as BSFL substrate.
3. Of the remaining 1,156 kg/ha of soybean, about 5 percent is consumed fresh (58 kg/ha) and another 5 percent is fed directly to livestock, leaving approximately 1,040 kg/ha of soybean for processing.
4. According to the director of a soybean processing plant in Ethiopia, soybean processing produces little waste. Soybean converts to soy cake, a product with two distinct markets: animal feed and human food. Because of its rich protein content, the primary market for soy cake is animal feed. The other significant market for soy cake is textured vegetable protein, which is used as a meat substitute for human consumption. Processing soy cake also yields soybean oil, commonly referred to as vegetable oil, which is used for cooking. Because of the utility of both soy cake and soybean oil, neither is considered a potential BSFL substrate. Only spoiled soy cake would be considered for BSFL but, according to the Ethiopian soybean processor, soy cake rarely spoils before it is used. However, there is some waste associated with soybean processing. Approximately 2 percent of incoming soybean is rejected upon inspection. These rejected beans, approximately 21 kg/ha of soybean, can be used as BSFL substrate.
5. Distribution waste and consumption waste for soybean are not considered for this model.

Groundnut

Figure 4.13 shows the food supply chain and associated waste stream for groundnut in Zimbabwe. The following analysis determines BSFL meal and frass production based on the five-year (2013–17) averages for total annual groundnut production (63,656 tons) and harvested area (168,043 ha). The numbered bullets below show the kilogram per hectare calculations for each type of groundnut-associated waste. These are then multiplied by the total area under cultivation (hectares) and divided by 1,000 (kg/ton) to determine the total tons per hectare per waste type. Table 4.14 summarizes the total groundnut-derived wastes recoverable and suitable as substrates for BSF breeding. Table 4.15 applies the 10 and 30 percent conversion rates to the wastes in table 4.14 to calculate the rate of converting groundnut wastes to BSFL meal and frass.

FIGURE 4.13 Groundnut Food Supply Chain and Annual Waste Stream, Zimbabwe

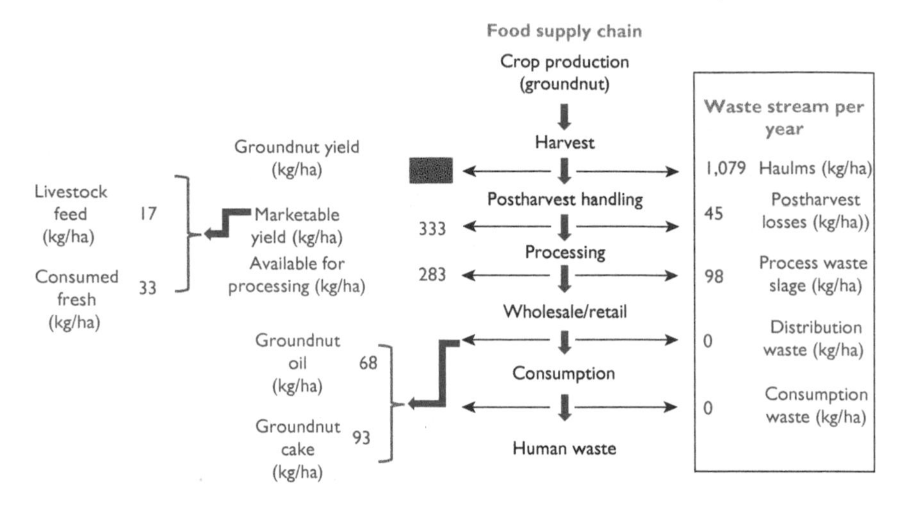

Source: Original figure for this publication.
Note: kg/ha = kilograms per hectare.

TABLE 4.14 Recoverable Groundnut-Derived Outputs for Black Soldier Fly Substrates, Zimbabwe

Waste/loss component	Recoverable output (kg/ha)	Total waste output (tons)
Haulms	1,079	181,311
Postharvest losses	45	7,634
Process waste	98	16,449

Source: Original table for this publication.
Note: kg/ha = kilograms per hectare.

1. Based on the five-year average of groundnut production and area cultivated, groundnut yield is 379 kg/ha and crop residue, or haulm, is 1,079 kg/ha. Like stover, haulm is good livestock fodder and a clean, abundant, readily available, and essentially no-cost substrate for BSFL.
2. Postharvest groundnut losses are approximately 12 percent of yield, or 45 kg/ha, which is a suitable amount for BSFL substrate, thus leaving 333 kg/ha of marketable groundnut.[12]
3. Humans consume approximately 10 percent, or 33 kg/ha, of the marketable groundnut, and livestock consume approximately 5 percent, or 17 kg/ha. This leaves 283 kg/ha of groundnut for processing.
4. Groundnut processing has two steps. The first step is hulling and deskinning. Groundnut hulls constitute approximately 25 percent and skins

TABLE 4.15 BSFL, Meal, and Frass Production from Groundnut Output Substrates, Zimbabwe

Waste/loss component	Total waste/loss (tons)	BSFL (tons)		BSFL meal (tons)		Frass (tons)	
		High	Low	High	Low	High	Low
Haulms	181,311	54,393	18,131	18,494	6,165	54,393	18,131
Postharvest losses	7,634	2,290	763	779	260	2,290	763
Process waste	16,449	4,935	1,645	1,678	559	4,935	1,645
TOTAL	205,394	61,618	20,539	20,950	6,983	61,618	20,539

Source: Original table for this publication.
Note: BSFL = black soldier fly larvae; high = 30 percent conversion; low = 10 percent conversion.

4 percent of groundnut yield by weight. Removing these elements results in a 29 percent loss of the initial 283 kg/ha of groundnut destined for processing. The remaining kernel amounts to 201 kg/ha, of which 4 percent is rejected, according to the director of a soybean processing plant in Ethiopia. This leaves 193 kg/ha of groundnut kernel for processing. The second processing step is groundnut crushing. This produces groundnut oil and meal (groundnut cake) and approximately 4 percent (nearly 8 kg/ha) of groundnut sludge. The sludge, hulls, skins, and rejected groundnut equal 98 kg/ha in total groundnut processing waste, which can be used as BSFL substrate. Like soybean cake, groundnut cake is used as a protein-rich livestock feed, and groundnut oil is also a marketable cooking product.

5. Distribution waste and consumption waste for groundnut production are not considered in this model.

Wheat

Figure 4.14 shows the food supply chain and associated waste stream for wheat in Zimbabwe. The following analysis determines BSFL meal and frass production based on the five-year (2013–17) averages for total annual wheat production (39,413 tons) and harvested area (19,423 ha). The numbered bullets below show the kilogram per hectare calculations for each type of wheat-associated waste. These are then multiplied by the total area under cultivation (hectares) and divided by 1,000 (kg/ton) to determine the total tons per hectare per waste type. Table 4.16 summarizes the total wheat-derived wastes recoverable and suitable as substrates for BSF breeding. Table 4.17 applies the 10 and 30 percent conversion rates to the wastes in table 4.16 to calculate the rate of converting wheat waste to BSFL meal and frass.

FIGURE 4.14 Wheat Food Supply Chain and Annual Waste Stream, Zimbabwe

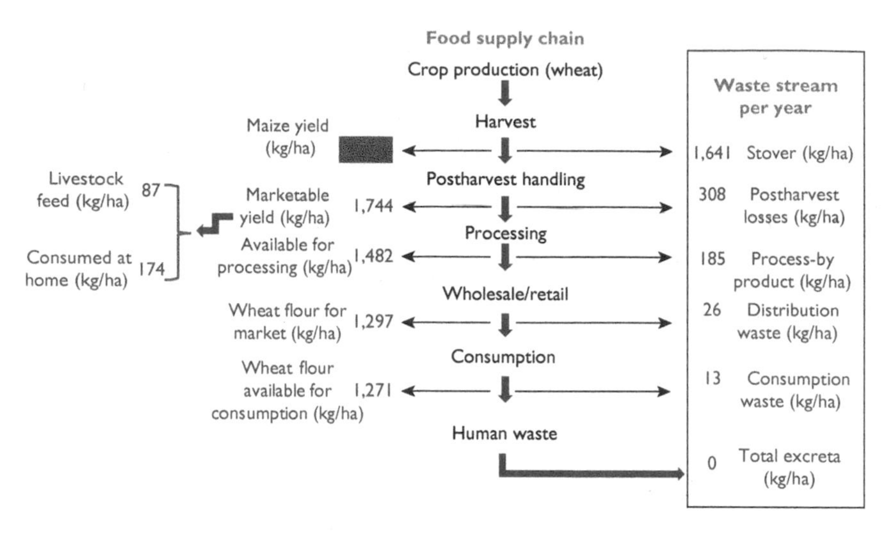

Source: Original figure for this publication.
Note: kg/ha = kilograms per hectare.

TABLE 4.16 Recoverable Wheat-Derived Outputs for Black Soldier Fly Substrates, Zimbabwe

Waste/loss component	Recoverable output (kg/ha)	Total waste output (tons)
Straw	1,641	31,874
Postharvest losses	308	5,976
Process waste	185	3,598
Consumption waste	13	247

Source: Original table for this publication.
Note: kg/ha = kilograms per hectare.

1. Based on the five-year average of wheat production and area cultivated, wheat yield is 2,051 kg/ha and crop residue, or straw, is 1,641 kg/ha. Wheat straw has low nutritional value but can still be used as animal fodder if nutrient supplements are added. However, as with other crop residues, straw can be a clean, abundant, readily available, and essentially no-cost substrate for BSFL.
2. Postharvest losses equal approximately 15 percent of wheat yield. This leaves 1,744 kg/ha of marketable wheat and 308 kg/ha of straw as a potential BSFL substrate.

TABLE 4.17 BSFL, Meal, and Frass Production from Wheat Output Substrates, Zimbabwe

Waste/loss component	Total waste/loss (tons)	BSFL (tons)		BSFL meal (tons)		Frass (tons)	
		High	Low	High	Low	High	Low
Straw	31,874	9,562	3,187	3,251	1,084	9,562	3,187
Postharvest losses	5,976	1,793	598	610	203	1,793	598
Process waste	3,598	1,079	360	367	122	1,079	360
Consumption waste	247	74	25	25	8	74	25
TOTAL	41,696	12,509	4,170	4,253	1,418	12,509	4,170

Source: Original table for this publication.
Note: BSFL = black soldier fly larvae; high = 30 percent conversion; low = 10 percent conversion.

3. Humans consume approximately 10 percent, or 174 kg/ha, of the marketable wheat, and livestock consume 5 percent, or 87 kg/ha, leaving 1,482 kg/ha of wheat for processing.
4. A wheat kernel comprises three parts: endosperm, which makes up approximately 83 percent of the kernel; bran, which makes up 14.5 percent of the kernel; and germ, which makes up 2.5 percent of the kernel (BAKERpedia 2020). The wheat milling process separates the bran and germ from the endosperm, which is used to make flour. Endosperm extraction ranges from 72 to 76 percent of the kernel's weight, but the model uses 74 percent. This means that endosperm extraction leaves approximately 9 percent of the kernel's weight as waste. This waste, along with the bran and germ by-products, results in 27 percent of the kernel's weight available for BSFL. Bran and germ are often burned or otherwise discarded despite both having nutritional value for humans and livestock. This model assumes that 20 percent of bran and germ will be available as BSFL substrate. Given these percentages, the total potential available wheat processing–derived substrate is calculated to be the following:
   - Endosperm waste: 9 percent of kernel weight => 0.09*1,482 kg/ha = 133 kg/ha
   - Germ: 2.5 percent of kernel weight => 0.029*1,482*0.2 = 9 kg/ha
   - Bran: 14.5 percent of kernel weight => 0.145*1,482*0.2 = 43 kg/ha
   - Total available for BSFL substrate: 185 kg/ha.
5. According to the FAO, consumption waste for cereals in Sub-Saharan Africa is roughly 1 percent. One percent of the approximately 1,297 kg/ha of milled wheat produced in processing equates to 13 kg/ha of consumption waste available as BSFL substrate (FAO 2011).

Economic and Social Benefits Associated with BSF Breeding

This subsection calculates the amount and value of protein and frass produced by BSFL from the various crop substrates and some of the environmental and employment benefits of BSF breeding. Table 4.18 summarizes the potential quantities of dry meal and frass derived from BSFL using wastes of the five crops as substrates. The table suggests that, on a per-hectare basis, sugarcane is the most favorable of the five crops for converting waste to BSFL and, eventually, to dry meal and frass. Maize provides the least waste that potentially can be used for BSF breeding. Per hectare, sugarcane produces more than 56 times the total waste as maize and greater than 18 times the total waste produced by wheat, the second leading crop in terms of per unit area waste production. The theoretical figures in table 4.18 assume full waste recovery at each step in the waste stream for all farms across Zimbabwe, but this is not realistic, nor is it realistic to assume that all farmers across the country will engage in BSF breeding. Even if a farmer captures 100 percent of the crop-related wastes for any or all of the five crops, the wastes can be used for purposes other than BSF breeding. Therefore, it would be unlikely that all the wastes would be used as BSFL substrate. As such, determining which crop has the most waste available for BSFL substrate would require knowing what percentages of each crop's wastes are diverted for other uses, but this may vary depending on location.

Protein

BSFL are an excellent source of protein for livestock and are profitable for farmers. The dry meal weight rendered from wet (live) larvae will be 34 percent of the weight of the wet larvae, or rather 34 percent wet larvae to dry meal conversion rate. The crude protein content of the dry meal is about 40 to 50 percent.[13] This compares favorably with the protein content of soy cake (47 to 49 percent) (Heuzé, Tran, and Kaushik 2020) and groundnut cake (45 to 60 percent) (Purohit and Rajyalakshmi 2011). Table 4.19 exhibits the protein demands of livestock in Zimbabwe, specifically, pigs, goats, chickens, and fish from aquaculture. Table 4.20 shows the crude protein derived from BSFL on the basis of the various crop residue substrates. BSFL dry meal meets between 6 percent (38,834 tons) and 17 percent (116,502 tons) of Zimbabwe's crude protein demand when comparing livestock's protein demand with BSFL-derived protein supply.[14] The model can also approximate the monetary value of BSFL crude protein by examining the international commodity price of soybean meal, which has similar protein content (48 percent) as BSFL. If the retail price of BSFL meal is assumed to be roughly the same as that of soybean meal (US$323 per ton),[15] the retail value of BSFL protein falls in the approximate range of US$12.55 million (for 38,834 tons) to US$37.65 million (for 166,502 tons) (table 4.21).

BSFL can cover protein demands for livestock. The BSFL crude protein totals used in this model assume the nationwide theoretical high values, which

TABLE 4.18 Output Summary of BSFL, Meal, and Frass from Five Key Crops, Zimbabwe

	Crop	Kilograms per hectare				Total cultivated area (ha)	National total (tons)			
		Total waste/loss	BSFL	BSFL meal	Frass		Total waste/loss	BSFL	BSFL meal	Frass
10% waste-to-larvae conversion	Maize	710	71	24	71	1,133,690	804,367	80,437	27,348	80,437
	Groundnut	1,222	122	42	122	168,043	205,394	20,539	6,983	20,539
	Soybean	1,557	156	53	156	43,713	68,080	6,808	2,315	6,808
	Sugarcane	39,551	3,955	1,345	3,955	43,890	1,735,913	173,591	59,021	173,591
	Wheat	2,147	215	73	215	19,423	41,696	4,170	1,418	4,170
	Total	**45,187**	**4,519**	**1,536**	**4,519**		**2,855,450**	**285,545**	**97,085**	**285,545**
30% waste-to-larvae conversion	Maize	710	213	72	213	1,133,690	804,367	241,310	82,045	241,310
	Groundnut	1,222	367	125	367	168,043	205,394	61,618	20,950	61,618
	Soybean	1,557	467	159	467	43,713	68,080	20,424	6,944	20,424
	Sugarcane	39,551	11,865	4,034	11,865	43,890	1,735,913	520,774	177,063	520,774
	Wheat	2,147	644	219	644	19,423	41,696	12,509	4,253	12,509
	Total	**45,187**	**13,556**	**4,609**	**13,556**		**2,855,450**	**856,635**	**291,256**	**856,635**

Source: Original table for this publication.
Note: BSFL = black soldier fly larvae; ha = hectares.

TABLE 4.19 Livestock Protein Demand, Zimbabwe

Livestock	Quantity	Total protein demand (tons)
Chickens	22,075,000	113,916
Pigs	251,758	6,878
Goats	4,895,043	564,056
Aquaculture (tons)	10,300	4,285
Total		**689,134**

Source: Original table for this publication.

TABLE 4.20 Total Dry and Crude Protein Converted from Five Key Crops, Zimbabwe

BSFL dry matter from	Total BSFL dry matter (tons)	Total crude protein (tons)	Total BSFL dry matter (tons)	Total crude protein (tons)
	10% conversion rate		30% conversion rate	
Maize	27,348	10,939	82,045	32,818
Groundnut	6,983	2,793	20,950	8,380
Soybean	2,315	926	6,944	2,778
Sugarcane	59,021	23,608	177,063	70,825
Wheat	1,418	567	4,253	1,701
TOTAL	**97,085**	**38,834**	**291,256**	**116,502**
BSFL protein surplus (deficit)		**(650,300)**		**(572,632)**
Crude protein/dry matter for BSFL (%)	*40*			
Total protein demand provided by BSFL (%)		**5.6**		**16.9**

Source: Original table for this publication.
Note: BSFL = black soldier fly larvae.

TABLE 4.21 Approximate Retail Value of BSFL Protein, Zimbabwe

Conversion rate (%)	BSFL meal (tons)	Value (US$, millions)
10	38,834	12.55
30	166,502	37.65

Source: Original table for this publication, using the price of soybean meal on September 25, 2019, per https://www.indexmundi.com/commodities/?commodity=soybean-meal.
Note: The table assumes the retail price of BSFL meal is approximately that of soybean meal. Much of this represents savings to the farmers who breed BSF since they are using their own on-farm wastes to produce this crude protein. BSFL = black soldier fly larvae.

are based on 100 percent recovery of crop wastes for substrate use. Actual protein totals will scale down according to the recovery level achievable for each crop substrate. Scaling is linear, meaning that if the overall recovery rate is 50 percent and all of that is used as substrate, then the resultant available crude protein will fall by 50 percent. For example, in table 4.20, the theoretical value (100 percent waste recovery) of total crude protein produced is 116,502 tons, assuming a 30 percent waste conversion factor. However, if only 50 percent of wastes are recovered and used as substrate, the resultant crude protein will be 50 percent of 116,502 tons, or 58,251 tons of crude protein, which would cover 8.5 percent of total protein demand (see table 4.20).

Frass for Biofertilizer

BSFL produces an abundant amount of high-quality biofertilizer, which is the BSFL's excrement (frass), mixed with the spent substrate. The amount of frass produced by BSFL from the various crop substrates is summarized in table 4.22. An individual insect larva can consume between 25 and 500 milligrams of organic matter per day depending on the larva's size, the type of substrate available, and the feeding conditions, which include air supply, temperature, and moisture levels (Makkar et al. 2014). Through the feeding process, BSFL release the substrate's nitrogen, phosphorus, and potassium (NPK). As such, the frass contains essential elements for plant growth and is used as

TABLE 4.22 Black Soldier Fly Larvae Frass Production, by Crop, Zimbabwe

	Crop	Residue (tons)	Post-harvest loss (tons)	Process waste (tons)	Consumption waste (tons)	Total (tons)
10% conversion	Maize	65,678	11,475	2,960	324	80,437
	Groundnut	18,131	763	1,645	—	20,539
	Soybean	6,395	322	91	—	6,808
	Sugarcane	54,273	3,618	115,700	—	173,591
	Wheat	3,187	598	360	25	4,170
	Total	**147,664**	**16,777**	**120,756**	**349**	**285,545**
30% conversion	Maize	197,033	34,425	8,881	971	241,310
	Groundnut	54,393	2,290	4,935	—	61,618
	Soybean	19,184	967	273	—	20,424
	Sugarcane	162,820	10,855	347,099	—	520,774
	Wheat	9,562	1,793	1,079	74	12,509
	Total	**442,992**	**50,330**	**362,267**	**1,046**	**856,635**

Source: Original table for this publication.
Note: — = not available.

crop fertilizer. The NPK value of the BSFL frass, expressed as a percentage of each nutrient,[16] varies depending on the producer. For example, online sales include BSFL frass NPK ratios as 5-3-2 and 3-2-4,[17] whereas in Thailand, as reported in table 3.11, the NPK ratio is 4.7-2.4-3. Regardless, the BSFL frass NPK concentration exceeds that of other common organic fertilizers such as chicken manure and cattle manure (table 4.23). The BSFL's exoskeleton, which can be retained as a component of BSFL frass-based biofertilizer,[18] contains the protein chitin, which when left in the soil aids plants' insect resistance and vitality.

BSFL frass has a potentially high market value. This can be approximated by using the international bulk price of NPK fertilizer, which is US$320 per ton, as a benchmark.[19] The resultant value of BSFL frass would be between approximately US$91.4 million (for 285,545 tons) and US$274.1 million (for 856,635 tons). BSFL frass is sold online, and the online retail price for a 1,650-pound tote (approximately 0.75 metric tons) of BSFL NPK 3-2-4 frass is US$3,000.[20] This equates to US$3,636 per metric ton. As such, the market value of frass produced by BSFL from various substrates of the five crops in Zimbabwe would range from US$1.038 billion to US$3.115 billion.

Employment

BSF-related activities are labor-intensive processes. A small-scale operator in Kenya reported needing 45 employees to produce three tons of BSFL dry meal per day. By contrast, large conventional animal feed mills require roughly 20 laborers to produce 100 tons of traditional feeds. Using the feed mill figures as a conservative employment metric, producing the BSFL, BSFL dry meal, and BSFL frass quantities calculated in this analysis would generate direct employment for 19,417 to 58,251 workers in Zimbabwe—specifically, workers in rural areas, where the majority of small-scale farms are located.

TABLE 4.23 NPK Values of Common Organic Fertilizers

Percent

Fertilizer source	N	P	K
Chicken with litter	3.7	1.4	2.8
Laying hen	1.9	1.3	2.0
Sheep	0.9	0.2	1.5
Rabbit	0.8	0.2	0.6
Beef (cattle)	0.6	0.1	0.7
Dry stack dairy	0.5	0.1	0.8
Horse	0.5	0.1	0.7

Source: Washington State University 2016.
Note: N = nitrogen; P = phosphorus; K = potassium.

The average real wage for medium-skilled workers in Zimbabwe in 2018 was approximately US$600 per month (WageIndicator Foundation 2020). As such, monthly wages for 19,417 to 58,251 workers would total approximately US$11,650,000 to US$34,951,000, respectively. The employment multiplier[21] for the feed industry is 2.76, meaning that for every direct job created, 2.76 indirect jobs will be produced.[22] Assuming that the employment structures of the traditional feed industry and the BSF-derived feed industry are parallel, multiplying the number of direct jobs by 2.76 suggests that the BSFL-derived feed industry would generate between 53,591 and 160,773 indirect jobs in Zimbabwe. Employment estimates for Zimbabwe are summarized in table 4.24.

Environment

BSFL production has fewer GHG emissions than other feed and fertilizer sources. Unlike synthetic fertilizer production and traditional livestock feed production, which produce crushed or pelletized grain, BSFL-derived biofertilizer and livestock feed production emit fewer GHG emissions and require very little energy (Parodi et al. 2020). Feed production for Zimbabwe's livestock population included in this analysis—pigs, goats, chickens, and fish from aquaculture—produces approximately 3,369,412 tons of CO_2-equivalent (CO_2-eq). BSFL meal production can meet between 6 and 17 percent of total livestock protein demand in Zimbabwe. This would reduce the GHG emissions from traditional feed production by between 190,465 and 571,396 tons of CO_2-eq. Likewise, producing synthetic NPK 15-15-15 fertilizer emits 0.76 kg of CO_2-eq per kilogram (Brentrup, Hoxha, and Christensen 2016). Reducing the production of synthetic fertilizer by the quantity of BSFL frass produced per the above analysis (285,454 to 856,635 tons) results in 217,014 to 651,042 tons fewer emissions of CO_2-eq.[23] As such, the overall reduction in GHG emissions from using BSFL frass and feed is between 407,479 and 1,222,438 tons of CO_2-eq. For comparison, the GHG emissions reduction in passenger vehicle equivalent would be 88,582 to 265,747 vehicles operating for a year (EPA 2018).

TABLE 4.24 Employment Generation Estimates Associated with Black Soldier Fly Breeding, Zimbabwe

Direct employment		Wage equivalent, direct employment total (US$)		Indirect employment		Total employment	
10%	30%	10%	30%	10%	30%	10%	30%
18,417	58,251	11,650,233	34,950,698	53,591	160,773	72,008	219,024

Source: Original table for this publication.
Note: 10% = low conversion rate; 30% = high conversion rate.

TABLE 4.25 GHG Emissions Reduction and Energy Savings from Using BSFL Meal and Frass Instead of Traditional Livestock Feed Production and Synthetic Fertilizers, Zimbabwe

Reduction/savings	10% conversion	30% conversion
GHG emissions reduction from BSFL meal (t CO_2-eq)	190,465	571,396
GHG emissions reduction from frass (t CO_2-eq)	217,014	651,042
Total GHG emissions reduction (t CO_2-eq)	407,479	1,222,438
• GHG emissions equivalent (passenger vehicles per year)	88,582	265,747
Energy savings from BSFL meal (MWh)	354,088	1,062,264
Energy savings from frass (MWh)	602,024	1,806,071
Total energy savings (MWh)	956,112	2,868,335
• Number of average homes in the northwestern United States that can be powered for one year by energy savings	87,316	261,948

Source: Original table for this publication.
Note: BSFL = black soldier fly larvae; CO_2-eq = carbon dioxide equivalent; GHG = greenhouse gas; MWh = megawatt-hours; t = tons.

BSFL feed and frass production require less energy than traditional livestock feed and synthetic fertilizer production. Replacing 6 to 17 percent of traditional animal feed production with BSFL meal production reduces energy consumption by 354,088 to 1,062,264 megawatt-hours (MWh). Replacing synthetic NPK fertilizer with an equivalent quantity of frass reduces energy needs by 602,024 to 1,806,071 MWh. Total energy savings from BSFL frass and meal production are 956,112 or 2,868,335 MWh, depending on the crop conversion rate. Table 4.25 summarizes these GHG emissions reductions and energy savings from BSFL-derived meal and frass production.

Extended Five-Crop Results Aggregated for Africa

The calculations in this section were extended to all of Africa. The five crops analyzed for Zimbabwe—maize, wheat, soybean, groundnut, and sugarcane—are pervasive throughout Africa. According to the FAO, maize and sugarcane, in particular, ranked second and third behind cassava as the most produced crops in Africa, by tonnage, from 2013 to 2018. Wheat consistently ranked among the top 10 in tons produced; groundnut ranked among the top 20; and soybean, while not as pervasive as the other four crops, ranked 45th in 2013 and consistently moved up in the rankings each year to 37th place in 2018. As such, the model uses the same five crops to model BSFL production and its impacts for all of Africa. Tables 4.26 to 4.29 report the aggregated results, again using the 2013–17 five-year average data for production and area harvested according to the 10 and 30 percent conversion rates.

TABLE 4.26 Crop Waste from Maize, Sugarcane, Groundnut, Soybean, and Wheat and the Amount of BSFL and Frass It Can Produce, Aggregates for All of Africa

Crop	Total crop-related waste (tons)	BSFL produced (tons)		BSFL frass produced (tons)	
		10%	30%	10%	30%
Maize	83,280,495	8,328,050	24,984,149	8,328,050	24,984,149
Groundnut	42,755,919	4,275,592	12,826,776	4,275,592	12,826,776
Soybean	3,265,735	326,573	979,720	326,573	979,720
Sugarcane	44,987,022	4,498,702	13,496,107	4,498,702	13,496,107
Wheat	27,736,047	2,773,605	8,320,814	2,773,605	8,320,814
Total	**202,025,218**	**20,202,522**	**60,607,565**	**20,202,522**	**60,607,565**

Source: Original table for this publication.
Note: 10% = low conversion rate; 30% = high conversion rate; BSFL = black soldier fly larvae.

TABLE 4.27 Livestock Protein Demand, Aggregates for All of Africa

Livestock/crop	Quantity	Total protein demand (tons)
Chickens	1,897,326,000	9,790,961
Pigs	37,377,009	1,021,140
Goats	426,257,039	49,117,599
Aquaculture (tons)	2,066,455	859,645
Total		**60,789,345**

BSFL dry matter from	Total BSFL dry matter (tons)	Total crude protein (tons)	Total BSFL dry matter (tons)	Total crude protein (tons)
	10% conversion rate		30% conversion rate	
Maize	2,831,537	1,132,615	8,494,611	3,397,844
Groundnut	1,453,701	581,480	4,361,104	1,744,441
Soybean	111,035	44,414	333,105	133,242
Sugarcane	1,529,559	611,823	4,588,676	1,835,470
Wheat	943,026	377,210	2,829,077	1,131,631
TOTAL	**6,868,857**	**2,747,543**	**20,606,572**	**8,242,629**

(Continued)

TABLE 4.27 Livestock Protein Demand, Aggregates for All of Africa *(Continued)*

BSFL dry matter from	Total BSFL dry matter (tons)	Total crude protein (tons)	Total BSFL dry matter (tons)	Total crude protein (tons)
	10% conversion rate		30% conversion rate	
BSFL protein surplus (deficit)		**(58,041,802)**		**(52,546,716)**
Crude protein/ dry matter for BSFL (%)	*40*			
Total protein demand provided by BSFL (%)		**4.5**		**13.6**

Source: Original table for this publication.
Note: BSFL = black soldier fly larvae.

TABLE 4.28 Employment Generation from Black Soldier Fly Production, for 10 and 30 Percent Conversion Rates, Aggregates for All of Africa

Direct employment		Indirect employment		Total employment	
10%	30%	10%	30%	10%	30%
1,373,771	4,121,314	3,791,609	11,374,828	5,165,381	15,496,142

Source: Original table for this publication.
Note: 10% = low conversion rate; 30% = high conversion rate.

TABLE 4.29 GHG Emissions Reduction and Energy Savings from Using BSFL Meal and Frass Instead of Other Meals and Organic Fertilizers, for 10 and 30 Percent Conversion Rates: Aggregates for All of Africa

Reduction/savings	10%	30%
GHG emissions reduction from BSFL meal (t CO_2-eq)	13,450,785	40,352,356
GHG emissions reduction from frass (t CO_2-eq)	15,353,917	46,061,750
Total GHG emissions reduction (t CO_2-eq)	28,804,702	86,414,106
• GHG emissions equivalent (passenger vehicles per year)	6,261,892	18,785,675
Energy savings from BSFL meal (MWh)	287,867	863,600
Energy savings from frass (MWh)	42,593,643	127,780,930
Total energy savings (MWh)	42,881,510	128,644,530
• Number of average homes in the northwestern United States that can be powered for one year by energy savings	3,916,120	11,748,359

Source: Original table for this publication.
Note: 10% = low conversion rate; 30% = high conversion rate; BSFL = black soldier fly larvae; CO_2-eq = carbon dioxide equivalent; GHG = greenhouse gas; MWh = megawatt-hours; t = tons.

ANNEX 4A. PROFILES OF POTENTIAL BENEFITS DERIVED FROM BLACK SOLDIER FLY IN 10 AFRICAN COUNTRIES

This chapter analyzed the results of using crop-related wastes and residues as substrate for black soldier fly larvae in Zimbabwe, along with the extension of that analysis for Africa as a whole. This annex extends the analysis to 10 additional countries in Africa individually, based on the dollar value of the agriculture sector in the respective countries. Selecting the countries entailed multiplying each country's percentage of gross domestic product attributed to agriculture by the total gross domestic product (table 4A.1). The 10 countries with the highest results were chosen for analysis.

In expanding the analysis to the additional 10 countries, the five crops associated with Zimbabwe did not appear consistently as key crops in the 10 countries. To compensate for the gaps, a sixth crop—cassava—was added to the analysis. However, only five crops were considered for each country. The

TABLE 4A.1 Countries Ranked by Value of Agriculture

Rank	Country	Agriculture in GDP (%)	GDP (US$, millions)	Ag value (US$, millions)
1	Nigeria	21.2	397,269.62	84,221.16
2	Kenya	34.2	87,908.26	30,064.62
3	Egypt, Arab Rep.	11.2	250,894.76	28,100.21
4	Ethiopia	31.2	84,355.60	26,318.95
5	Algeria	12.0	173,757.95	20,850.95
6	Tanzania	28.7	58,001.20	16,646.34
7	Morocco	12.3	117,921.39	14,504.33
8	Sudan	31.5	40,851.54	12,868.24
9	Ghana	18.3	65,556.46	11,996.83
10	Angola	10.0	105,750.99	10,575.10
11	Uganda	24.2	43,007.05	10,407.71
12	Congo, Dem. Rep.	19.1	47,227.54	9,020.46
13	Côte d'Ivoire	19.8	43,007.05	8,515.40
14	South Africa	2.2	368,288.94	8,102.36
15	Mali	38.7	17,163.43	6,642.25
	Zimbabwe	12.1	31,000.52	3,751.06

Source: Original table for this publication, using data from the World Bank.
Note: Data are for 2018 except in the cases of Angola and Tanzania, for which 2017 data are used. GDP = gross domestic product.

crops by country are based on the production of each crop in the respective country. The five highest ranking crops in terms of production tonnage were chosen for each respective country. Even with the addition of cassava, two countries—Algeria and Sudan—do not have five of the six selected crops as key crops, or they are not listed in the FAO database as being produced in that country. For Algeria, the FAO includes wheat, groundnut, and maize; thus, cassava, sugarcane, and soybean are not listed as key crops for Algeria. In Algeria, even the production of groundnut and wheat is relatively low, at less than 10,000 tons each (5,337 and 2,586 tons, respectively). Sudan produces four of the six crops listed—sugarcane, groundnut, wheat, and maize (there are no data for cassava or soybean).

The results for each country are presented in the following subsections.

NIGERIA

TABLE 4A.2 Crop Waste, BSFL, and Frass, Nigeria

Crop	Total crop-related waste (tons)	BSFL produced (tons)		BSFL frass produced (tons)	
		10%	30%	10%	30%
Maize	11,098,536	1,109,854	3,329,561	1,109,854	3,329,561
Groundnut	9,918,854	991,885	2,975,656	991,885	2,975,656
Soybean	779,951	77,995	233,985	77,995	233,985
Sugarcane	684,947	68,495	205,484	68,495	205,484
Cassava	30,377,100	3,037,710	9,113,130	3,037,710	9,113,130
Total	**52,859,388**	**5,285,939**	**15,857,816**	**5,285,939**	**15,857,816**

Source: Original table for this publication.
Note: 10% = low conversion rate; 30% = high conversion rate; BSFL = black soldier fly larvae.

TABLE 4A.3 Livestock Protein Demand, Nigeria

Livestock/crop	Quantity	Total protein demand (tons)
Chickens	143,232,000	739,134
Pigs	78,049,310	2,132,307
Goats	7,498,342	864,034
Aquaculture (tons)	296,071	123,166
Total		**3,858,641**

(Continued)

TABLE 4A.3 Livestock Protein Demand, Nigeria *(Continued)*

BSFL dry matter from	Total BSFL dry matter (tons)	Total crude protein (tons)	Total BSFL dry matter (tons)	Total crude protein (tons)
	10% conversion rate		30% conversion rate	
Maize	377,350	150,940	1,132,051	452,820
Groundnut	337,241	134,896	1,011,723	404,689
Soybean	26,518	10,607	79,555	31,822
Sugarcane	23,288	9,315	69,865	27,946
Cassava	1,032,821	413,129	3,098,464	1,239,386
TOTAL	**1,797,219**	**718,888**	**5,391,658**	**2,156,663**
BSFL protein surplus (deficit)		**(3,139,753)**		**(1,701,978)**
Crude protein/dry matter for BSFL (%)	*40*			
Total protein demand provided by BSFL (%)		**18.6**		**55.9**

Source: Original table for this publication.
Note: BSFL = black soldier fly larvae.

TABLE 4A.4 Employment Generation, Nigeria

Direct employment		Indirect employment		Total employment	
10%	30%	10%	30%	10%	30%
359,444	1,078,332	992,065	2,976,195	1,351,509	4,054,526

Source: Original table for this publication.
Note: 10% = low conversion rate; 30% = high conversion rate.

TABLE 4A.5 GHG Emissions Reduction and Energy Savings Realized through BSFL-Derived Meal and Frass, Nigeria

Reduction/savings	10%	30%
GHG emissions reduction from BSFL meal (t CO_2-eq)	5,328,436	15,985,308
GHG emissions reduction from frass (t CO_2-eq)	4,017,313	12,051,940
Total GHG emissions reduction (t CO_2-eq)	9,345,749	28,037,248
• GHG emissions equivalent (passenger vehicles per year)	2,031,685	6,095,054
Energy savings from BSFL meal (MWh)	1,208,040	3,624,119
Energy savings from frass (MWh)	11,144,519	33,433,557
Total energy savings (MWh)	12,352,559	37,057,676
• Number of average homes in the northwestern United States that can be powered for one year by energy savings	1,128,088	3,384,263

Source: Original table for this publication.
Note: 10% = low conversion rate; 30% = high conversion rate; BSFL = black soldier fly larvae; CO_2-eq = carbon dioxide equivalent; GHG = greenhouse gas; MWh = megawatt-hours; t = tons.

KENYA

TABLE 4A.6 Crop Waste, BSFL, and Frass, Kenya

Crop	Total crop-related waste (tons)	BSFL produced (tons)		BSFL frass produced (tons)	
		10%	30%	10%	30%
Maize	3,927,676	392,768	1,178,303	392,768	1,178,303
Groundnut	124,286	12,429	37,286	12,429	37,286
Sugarcane	3,075,410	307,541	922,623	307,541	922,623
Wheat	269,834	26,983	80,950	26,983	80,950
Cassava	463,110	46,311	138,933	46,311	138,933
Total	**7,860,317**	**786,032**	**2,358,095**	**786,032**	**2,358,095**

Source: Original table for this publication.
Note: 10% = low conversion rate; 30% = high conversion rate; BSFL = black soldier fly larvae.

TABLE 4A.7 Livestock Protein Demand, Kenya

Livestock/crop	Quantity	Total protein demand (tons)		
Chickens	48,125,000	248,344		
Pigs	25,684,489	701,700		
Goats	554,301	63,872		
Aquaculture (tons)	12,360	5,142		
Total		**1,019,058**		
BSFL dry matter from	**Total BSFL dry matter (tons)**	**Total crude protein (tons)**	**Total BSFL dry matter (tons)**	**Total crude protein (tons)**
	10% conversion rate		**30% conversion rate**	
Maize	133,541	53,416	400,623	160,249
Groundnut	4,226	1,690	12,677	5,071
Sugarcane	104,564	41,826	313,692	125,477
Wheat	9,174	3,670	27,523	11,009
Cassava	15,746	6,298	47,237	18,895
TOTAL	**267,251**	**106,900**	**801,752**	**320,701**
BSFL protein surplus (deficit)		**(912,158)**		**(698,357)**
Crude protein/dry matter for BSFL (%)	*40*			
Total protein demand provided by BSFL (%)		**10.5**		**31.5**

Source: Original table for this publication.
Note: BSFL = black soldier fly larvae.

TABLE 4A.8 Employment Generation, Kenya

Direct employment		Indirect employment		Total employment	
10%	30%	10%	30%	10%	30%
53,450	160,350	147,522	442,567	200,973	602,918

Source: Original table for this publication.
Note: 10% = low conversion rate; 30% = high conversion rate.

TABLE 4A.9 GHG Emissions Reduction and Energy Savings Realized through BSFL-Derived Meal and Frass, Kenya

Reduction/savings	10%	30%
GHG emissions reduction from BSFL meal (t CO_2-eq)	858,867	2,576,600
GHG emissions reduction from frass (t CO_2-eq)	593,959	1,781,878
Total GHG emissions reduction (t CO_2-eq)	1,452,826	4,358,478
• GHG emissions equivalent (passenger vehicles per year)	315,832	947,495
Energy savings from BSFL meal (MWh)	664,648	1,993,945
Energy savings from frass (MWh)	1,647,716	4,943,147
Total energy savings (MWh)	2,312,364	6,937,092
• Number of average homes in the northwestern United States that can be powered for one year by energy savings	211,175	633,524

Source: Original table for this publication.
Note: 10% = low conversion rate; 30% = high conversion rate; BSFL = black soldier fly larvae; CO_2-eq = carbon dioxide equivalent; GHG = greenhouse gas; MWh = megawatt-hours; t = tons.

ARAB REPUBLIC OF EGYPT

TABLE 4A.10 Crop Waste, BSFL, and Frass, Arab Republic of Egypt

Crop	Total crop-related waste (tons)	BSFL produced (tons)		BSFL frass produced (tons)	
		10%	30%	10%	30%
Maize	8,758,878	875,888	2,627,663	875,888	2,627,663
Groundnut	325,919	32,592	97,776	32,592	97,776
Soybean	51,124	5,112	15,337	5,112	15,337
Sugarcane	7,549,131	754,913	2,264,739	754,913	2,264,739
Wheat	9,562,319	956,232	2,868,696	956,232	2,868,696
Total	**26,247,372**	**2,624,737**	**7,874,212**	**2,624,737**	**7,874,212**

Source: Original table for this publication.

Note: 10% = low conversion rate; 30% = high conversion rate; BSFL = black soldier fly larvae.

TABLE 4A.11 Livestock Protein Demand, Arab Republic of Egypt

Livestock/crop	Quantity	Total protein demand (tons)		
Chickens	144,797,000	747,210		
Pigs	3,973,692	108,561		
Goats	9,766	1,125		
Aquaculture (tons)	1,451,706	603,910		
Total		**1,460,807**		
BSFL dry matter from	**Total BSFL dry matter (tons)**	**Total crude protein (tons)**	**Total BSFL dry matter (tons)**	**Total crude protein (tons)**
	10% conversion rate		**30% conversion rate**	
Maize	297,802	119,121	893,406	357,362
Groundnut	11,081	4,433	33,244	13,298
Soybean	1,738	695	5,215	2,086
Sugarcane	256,670	102,668	770,011	308,005
Wheat	325,119	130,048	975,357	390,143
TOTAL	**892,411**	**356,964**	**2,677,232**	**1,070,893**
BSFL protein surplus (deficit)		**(1,103,842)**		**(389,914)**
Crude protein/dry matter for BSFL (%)	*40*			
Total protein demand provided by BSFL (%)		**24.4**		**73.3**

Source: Original table for this publication.

Note: BSFL = black soldier fly larvae.

TABLE 4A.12 Employment Generation, Arab Republic of Egypt

Direct employment		Indirect employment		Total employment	
10%	30%	10%	30%	10%	30%
178,482	535,446	492,611	1,477,832	671,093	2,013,278

Source: Original table for this publication.
Note: 10% = low conversion rate; 30% = high conversion rate.

TABLE 4A.13 GHG Emissions Reduction and Energy Savings Realized through BSFL-Derived Meal and Frass, Arab Republic of Egypt

Reduction/savings	10%	30%
GHG emissions reduction from BSFL meal (t CO_2-eq)	512,791	1,538,373
GHG emissions reduction from frass (t CO_2-eq)	1,994,800	5,984,401
Total GHG emissions reduction (t CO_2-eq)	2,507,591	7,522,773
• GHG emissions equivalent (passenger vehicles per year)	545,129	1,635,386
Energy savings from BSFL meal (MWh)	1,533,244	4,599,731
Energy savings from frass (MWh)	5,533,820	16,601,460
Total energy savings (MWh)	7,067,064	21,201,191
• Number of average homes in the northwestern United States that can be powered for one year by energy savings	645,394	1,936,182

Source: Original table for this publication.
Note: 10% = low conversion rate; 30% = high conversion rate; BSFL = black soldier fly larvae; CO_2-eq = carbon dioxide equivalent; GHG = greenhouse gas; MWh = megawatt-hours; t = tons.

ETHIOPIA

TABLE 4A.14 Crop Waste, BSFL, and Frass, Ethiopia

Crop	Total crop-related waste (tons)	BSFL produced (tons)		BSFL frass produced (tons)	
		10%	30%	10%	30%
Maize	8,157,353	815,735	2,447,206	815,735	2,447,206
Groundnut	389,155	38,915	116,746	38,915	116,746
Soybean	97,244	9,724	29,173	9,724	29,173
Sugarcane	801,666	80,167	240,500	80,167	240,500
Wheat	4,630,218	463,022	1,389,065	463,022	1,389,065
Total	**14,075,635**	**1,407,563**	**4,222,690**	**1,407,563**	**4,222,690**

Source: Original table for this publication.

Note: 10% = low conversion rate; 30% = high conversion rate; BSFL = black soldier fly larvae.

TABLE 4A.15 Livestock Protein Demand, Ethiopia

Livestock/crop	Quantity	Total protein demand (tons)		
Chickens	59,158,000	305,279		
Pigs	30,747,916	840,033		
Goats	35,388	4,078		
Aquaculture (tons)	126	52		
Total		**1,149,442**		
BSFL dry matter from	**Total BSFL dry matter (tons)**	**Total crude protein (tons)**	**Total BSFL dry matter (tons)**	**Total crude protein (tons)**
	10% conversion rate		**30% conversion rate**	
Maize	277,350	110,940	832,050	332,820
Groundnut	13,231	5,293	39,694	15,878
Soybean	3,306	1,323	9,919	3,968
Sugarcane	27,257	10,903	81,770	32,708
Wheat	157,427	62,971	472,282	188,913
TOTAL	**478,572**	**191,429**	**1,435,715**	**574,286**
BSFL protein surplus (deficit)		**(958,014)**		**(575,156)**
Crude protein/dry matter for BSFL (%)	*40*			
Total protein demand provided by BSFL (%)		**16.7**		**50.0**

Source: Original table for this publication.

Note: BSFL = black soldier fly larvae.

TABLE 4A.16 Employment Generation, Ethiopia

Direct employment		Indirect employment		Total employment	
10%	30%	10%	30%	10%	30%
95,714	287,143	264,172	792,515	359,886	1,079,658

Source: Original table for this publication.
Note: 10% = low conversion rate; 30% = high conversion rate.

TABLE 4A.17 GHG Emissions Reduction and Energy Savings Realized through BSFL-Derived Meal and Frass, Ethiopia

Reduction/savings	10%	30%
GHG emissions reduction from BSFL meal (t CO_2-eq)	1,566,546	4,699,639
GHG emissions reduction from frass (t CO_2-eq)	1,069,748	3,209,245
Total GHG emissions reduction (t CO_2-eq)	2,636,295	7,908,884
• GHG emissions equivalent (passenger vehicles per year)	573,108	1,719,323
Energy savings from BSFL meal (MWh)	1,057,579	3,172,738
Energy savings from frass (MWh)	2,967,613	8,902,838
Total energy savings (MWh)	4,025,192	12,075,576
• Number of average homes in the northwestern United States that can be powered for one year by energy savings	367,597	1,102,792

Source: Original table for this publication.
Note: 10% = low conversion rate; 30% = high conversion rate; BSFL = black soldier fly larvae; CO_2-eq = carbon dioxide equivalent; GHG = greenhouse gas; MWh = megawatt-hours; t = tons.

ALGERIA

TABLE 4A.18 Crop Waste, BSFL, and Frass, Algeria

Crop	Total crop-related waste (tons)	BSFL produced (tons)		BSFL frass produced (tons)	
		10%	30%	10%	30%
Maize	3,363	336	1,009	336	1,009
Groundnut	16,154	1,615	4,846	1,615	4,846
Wheat	2,818,456	281,846	845,537	281,846	845,537
Total	**2,837,973**	**283,797**	**851,392**	**283,797**	**851,392**

Source: Original table for this publication.
Note: 10% = low conversion rate; 30% = high conversion rate; BSFL = black soldier fly larvae.

TABLE 4A.19 Livestock Protein Demand, Algeria

Livestock/crop	Quantity	Total protein demand (tons)		
Chickens	135,018,000	696,747		
Pigs	5,007,894	136,816		
Goats	4,789	552		
Aquaculture (tons)	1,408	586		
Total		**834,700**		
BSFL dry matter from	**Total BSFL dry matter (tons)**	**Total crude protein (tons)**	**Total BSFL dry matter (tons)**	**Total crude protein (tons)**
	10% conversion rate		**30% conversion rate**	
Maize	114	46	343	137
Groundnut	549	220	1,648	659
Wheat	95,828	38,331	287,483	114,993
TOTAL	**96,491**	**38,596**	**289,473**	**115,789**
BSFL protein surplus (deficit)		**(796,104)**		**(718,911)**
Crude protein/dry matter for BSFL (%)	*40*			
Total protein demand provided by BSFL (%)		**4.6**		**13.9**

Source: Original table for this publication.
Note: BSFL = black soldier fly larvae.

TABLE 4A.20 Employment Generation, Algeria

Direct employment		Indirect employment		Total employment	
10%	30%	10%	30%	10%	30%
19,298	57,895	53,263	159,789	72,561	217,684

Source: Original table for this publication.
Note: 10% = low conversion rate; 30% = high conversion rate.

TABLE 4A.21 GHG Emissions Reduction and Energy Savings Realized through BSFL-Derived Meal and Frass, Algeria

Reduction/savings	10%	30%
GHG emissions reduction from BSFL meal (t CO_2-eq)	107,994	323,983
GHG emissions reduction from frass (t CO_2-eq)	215,686	647,058
Total GHG emissions reduction (t CO_2-eq)	323,680	971,040

(Continued)

TABLE 4A.21 GHG Emissions Reduction and Energy Savings Realized through BSFL-Derived Meal and Frass, Algeria *(Continued)*

Reduction/savings	10%	30%
• GHG emissions equivalent (passenger vehicles per year)	70,365	211,096
Energy savings from BSFL meal (MWh)	290,268	870,803
Energy savings from frass (MWh)	598,339	1,795,018
Total energy savings (MWh)	888,607	2,665,820
• Number of average homes in the northwestern United States that can be powered for one year by energy savings	81,151	243,454

Source: Original table for this publication.
Note: 10% = low conversion rate; 30% = high conversion rate; BSFL = black soldier fly larvae; CO_2-eq = carbon dioxide equivalent; GHG = greenhouse gas; MWh = megawatt-hours; t = tons.

TANZANIA

TABLE 4A.22 Crop Waste, BSFL, and Frass, Tanzania

Crop	Total crop-related waste (tons)	BSFL produced (tons)		BSFL frass produced (tons)	
		10%	30%	10%	30%
Maize	6,495,806	649,581	1,948,742	649,581	1,948,742
Groundnut	4,390,920	439,092	1,317,276	439,092	1,317,276
Soybean	7,477	748	2,243	748	2,243
Wheat	112,898	11,290	33,869	11,290	33,869
Cassava	2,868,877	286,888	860,663	286,888	860,663
Total	**13,875,977**	**1,387,598**	**4,162,793**	**1,387,598**	**4,162,793**

Source: Original table for this publication.
Note: 10% = low conversion rate; 30% = high conversion rate; BSFL = black soldier fly larvae.

TABLE 4A.23 Livestock Protein Demand, Tanzania

Livestock/crop	Quantity	Total protein demand (tons)
Chickens	37,492,000	193,474
Pigs	17,938,696	490,085
Goats	518,881	59,791
Aquaculture (tons)	11,350	4,722
Total		**748,071**

(Continued)

TABLE 4A.23 Livestock Protein Demand, Tanzania *(Continued)*

BSFL dry matter from	Total BSFL dry matter (tons)	Total crude protein (tons)	Total BSFL dry matter (tons)	Total crude protein (tons)
	10% conversion rate		30% conversion rate	
Maize	220,857	88,343	662,572	265,029
Groundnut	149,291	59,717	447,874	179,150
Soybean	254	102	763	305
Wheat	3,839	1,535	11,516	4,606
Cassava	97,542	39,017	292,625	117,050
TOTAL	**471,783**	**188,713**	**1,415,350**	**566,140**
BSFL protein surplus (deficit)		**(559,358)**		**(181,931)**
Crude protein/ dry matter for BSFL (%)	*40*			
Total protein demand provided by BSFL (%)		**25.2**		**75.7**

Source: Original table for this publication.
Note: BSFL = black soldier fly larvae.

TABLE 4A.24 Employment Generation, Tanzania

Direct employment		Indirect employment		Total employment	
10%	30%	10%	30%	10%	30%
94,357	283,070	260,424	781,273	354,781	1,064,343

Source: Original table for this publication.
Note: 10% = low conversion rate; 30% = high conversion rate.

TABLE 4A.25 GHG Emissions Reduction and Energy Savings Realized through BSFL-Derived Meal and Frass, Tanzania

Reduction/savings	10%	30%
GHG emissions reduction from BSFL meal (t CO_2-eq)	1,470,352	4,411,057
GHG emissions reduction from frass (t CO_2-eq)	1,053,596	3,160,789
Total GHG emissions reduction (t CO_2-eq)	2,523,949	7,571,847
• GHG emissions equivalent (passenger vehicles per year)	548,645	1,646,054
Energy savings from BSFL meal (MWh)	1,592,818	4,778,454
Energy savings from frass (MWh)	2,922,806	8,768,417
Total energy savings (MWh)	4,515,623	13,546,870

(Continued)

TABLE 4A.25 GHG Emissions Reduction and Energy Savings Realized through BSFL-Derived Meal and Frass, Tanzania *(Continued)*

Reduction/savings	10%	30%
• Number of average homes in the northwestern United States that can be powered for one year by energy savings	412,386	1,237,157

Source: Original table for this publication.
Note: 10% = low conversion rate; 30% = high conversion rate; BSFL = black soldier fly larvae; CO_2-eq = carbon dioxide equivalent; GHG = greenhouse gas; MWh = megawatt-hours; t = tons.

MOROCCO

TABLE 4A.26 Crop Waste, BSFL, and Frass, Morocco

Crop	Total crop-related waste (tons)	BSFL produced (tons)		BSFL frass produced (tons)	
		10%	30%	10%	30%
Maize	123,702	12,370	37,111	12,370	37,111
Groundnut	118,333	11,833	35,500	11,833	35,500
Soybean	1,267	127	380	127	380
Sugarcane	218,159	21,816	65,448	21,816	65,448
Wheat	6,114,309	611,431	1,834,293	611,431	1,834,293
Total	**6,575,770**	**657,577**	**1,972,731**	**657,577**	**1,972,731**

Source: Original table for this publication.
Note: 10% = low conversion rate; 30% = high conversion rate; BSFL = black soldier fly larvae.

TABLE 4A.27 Livestock Protein Demand, Morocco

Livestock/crop	Quantity	Total protein demand (tons)
Chickens	200,493,000	1,034,624
Pigs	5,205,000	142,201
Goats	8,006	923
Aquaculture (tons)	787	327
Total		**1,178,075**

(Continued)

TABLE 4A.27 Livestock Protein Demand, Morocco *(Continued)*

BSFL dry matter from	Total BSFL dry matter (tons)	Total crude protein (tons)	Total BSFL dry matter (tons)	Total crude protein (tons)
	10% conversion rate		30% conversion rate	
Maize	4,206	1,682	12,618	5,047
Groundnut	4,023	1,609	12,070	4,828
Soybean	43	17	129	52
Sugarcane	7,417	2,967	22,252	8,901
Wheat	207,887	83,155	623,660	249,464
TOTAL	**223,576**	**89,430**	**670,729**	**268,291**
BSFL protein surplus (deficit)		**(1,088,644)**		**(909,783)**
Crude protein/dry matter for BSFL (%)	*40*			
Total protein demand provided by BSFL (%)		**7.6**		**22.8**

Source: Original table for this publication.
Note: BSFL = black soldier fly larvae.

TABLE 4A.28 Employment Generation, Morocco

Direct employment		Indirect employment		Total employment	
10%	30%	10%	30%	10%	30%
44,715	134,146	123,414	370,242	168,129	504,388

Source: Original table for this publication.
Note: 10% = low conversion rate; 30% = high conversion rate.

TABLE 4A.29 GHG Emissions Reduction and Energy Savings Realized through BSFL-Derived Meal and Frass, Morocco

Reduction/savings	10%	30%
GHG emissions reduction from BSFL meal (t CO_2-eq)	213,644	640,931
GHG emissions reduction from frass (t CO_2-eq)	499,759	1,499,276
Total GHG emissions reduction (t CO_2-eq)	713,402	2,140,207
• GHG emissions equivalent (passenger vehicles per year)	155,087	465,262
Energy savings from BSFL meal (MWh)	476,577	1,429,731

(Continued)

TABLE 4A.29 GHG Emissions Reduction and Energy Savings Realized through BSFL-Derived Meal and Frass, Morocco *(Continued)*

Reduction/savings	10%	30%
Energy savings from frass (MWh)	1,386,391	4,159,174
Total energy savings (MWh)	1,862,968	5,588,905
• Number of average homes in the northwestern United States that can be powered for one year by energy savings	170,134	510,402

Source: Original table for this publication.
Note: 10% = low conversion rate; 30% = high conversion rate; BSFL = black soldier fly larvae; CO_2-eq = carbon dioxide equivalent; GHG = greenhouse gas; MWh = megawatt-hours; t = tons.

SUDAN

TABLE 4A.30 Crop Waste, BSFL, and Frass, Sudan

Crop	Total crop-related waste (tons)	BSFL produced (tons)		BSFL frass produced (tons)	
		10%	30%	10%	30%
Maize	52,739	5,274	15,822	5,274	15,822
Groundnut	5,232,400	523,240	1,569,720	523,240	1,569,720
Sugarcane	2,832,691	283,269	849,807	283,269	849,807
Wheat	512,648	51,265	153,794	51,265	153,794
Total	**8,630,478**	**863,048**	**2,589,143**	**863,048**	**2,589,143**

Source: Original table for this publication.
Note: 10% = low conversion rate; 30% = high conversion rate; BSFL = black soldier fly larvae.

TABLE 4A.31 Livestock Protein Demand, Sudan

Livestock/crop	Quantity	Total protein demand (tons)
Chickens	48,584,000	250,713
Pigs	–	–
Goats	31,659,000	3,648,067
Aquaculture (tons)	9,000	3,744
Total		**3,902,523**

(Continued)

TABLE 4A.31 Livestock Protein Demand, Sudan *(Continued)*

BSFL dry matter from	Total BSFL dry matter (tons)	Total crude protein (tons)	Total BSFL dry matter (tons)	Total crude protein (tons)
	10% conversion rate		30% conversion rate	
Maize	1,793	717	5,379	2,152
Groundnut	177,902	71,161	533,705	213,482
Sugarcane	96,311	38,525	288,934	115,574
Wheat	17,430	6,972	52,290	20,916
TOTAL	**293,436**	**117,374**	**880,309**	**352,123**
BSFL protein surplus (deficit)		**(3,785,149)**		**(3,550,400)**
Crude protein/dry matter for BSFL (%)	*40*			
Total protein demand provided by BSFL (%)		**3.0**		**9.0**

Source: Original table for this publication.
Note: BSFL = black soldier fly larvae. — = not available.

TABLE 4A.32 Employment Generation, Sudan

Direct employment		Indirect employment		Total employment	
10%	30%	10%	30%	10%	30%
58,687	176,062	161,977	485,930	220,664	661,992

Source: Original table for this publication.
Note: 10% = low conversion rate; 30% = high conversion rate.

TABLE 4A.33 GHG Emissions Reduction and Energy Savings Realized through BSFL-Derived Meal and Frass, Sudan

Reduction/savings	10%	30%
GHG emissions reduction from BSFL meal (t CO_2-eq)	622,625	1,867,876
GHG emissions reduction from frass (t CO_2-eq)	655,916	1,967,749
Total GHG emissions reduction (t CO_2-eq)	1,278,542	3,835,625
• GHG emissions equivalent (passenger vehicles per year)	277,944	833,831
Energy savings from BSFL meal (MWh)	188,377	565,131
Energy savings from frass (MWh)	1,819,592	5,458,776
Total energy savings (MWh)	2,007,969	6,023,908
• Number of average homes in the northwestern United States that can be powered for one year by energy savings	183,376	550,129

Source: Original table for this publication.
Note: 10% = low conversion rate; 30% = high conversion rate; BSFL = black soldier fly larvae; CO_2-eq = carbon dioxide equivalent; GHG = greenhouse gas; MWh = megawatt-hours; t = tons.

GHANA

TABLE 4A.34 Crop Waste, BSFL, and Frass, Ghana

Crop	Total crop-related waste (tons)	BSFL produced (tons)		BSFL frass produced (tons)	
		10%	30%	10%	30%
Maize	1,956,206	195,621	586,862	195,621	586,862
Groundnut	1,363,582	136,358	409,074	136,358	409,074
Soybean	186,438	18,644	55,931	18,644	55,931
Sugarcane	72,191	7,219	21,657	7,219	21,657
Cassava	9,486,125	948,613	2,845,838	948,613	2,845,838
Total	**13,064,542**	**1,306,454**	**3,919,363**	**1,306,454**	**3,919,363**

Source: Original table for this publication.
Note: 10% = low conversion rate; 30% = high conversion rate; BSFL = black soldier fly larvae.

TABLE 4A.35 Livestock Protein Demand, Ghana

Livestock/crop	Quantity	Total protein demand (tons)		
Chickens	77,443,000	399,637		
Pigs	6,400,000	174,848		
Goats	819,843	94,471		
Aquaculture (tons)	57,405	23,880		
Total		**692,836**		
BSFL dry matter from	**Total BSFL dry matter (tons)**	**Total crude protein (tons)**	**Total BSFL dry matter (tons)**	**Total crude protein (tons)**
	10% conversion rate		**30% conversion rate**	
Maize	66,511	26,604	199,533	79,813
Groundnut	46,362	18,545	139,085	55,634
Soybean	6,339	2,536	19,017	7,607
Sugarcane	2,455	982	7,364	2,945
Cassava	322,528	129,011	967,585	387,034
TOTAL	**444,194**	**177,678**	**1,332,583**	**533,033**
BSFL protein surplus (deficit)		**(515,158)**		**(159,803)**
Crude protein/dry matter for BSFL (%)	*40*			
Total protein demand provided by BSFL (%)		**25.6**		**76.9**

Source: Original table for this publication.
Note: BSFL = black soldier fly larvae.

TABLE 4A.36 Employment Generation, Ghana

Direct employment		Indirect employment		Total employment	
10%	30%	10%	30%	10%	30%
88,839	266,517	245,195	735,586	334,034	1,002,103

Source: Original table for this publication.
Note: 10% = low conversion rate; 30% = high conversion rate.

TABLE 4A.37 GHG Emissions Reduction and Energy Savings Realized through BSFL-Derived Meal and Frass, Ghana

Reduction/savings	10%	30%
GHG emissions reduction from BSFL meal (t CO_2-eq)	743,588	2,230,764
GHG emissions reduction from frass (t CO_2-eq)	992,905	2,978,716
Total GHG emissions reduction (t CO_2-eq)	1,736,493	5,209,480
• GHG emissions equivalent (passenger vehicles per year)	377,499	1,132,496
Energy savings from BSFL meal (MWh)	1,610,857	4,832,571
Energy savings from frass (MWh)	2,754,441	8,263,322
Total energy savings (MWh)	4,365,298	13,095,893
• Number of average homes in the northwestern United States that can be powered for one year by energy savings	398,657	1,195,972

Source: Original table for this publication.
Note: 10% = low conversion rate; 30% = high conversion rate; BSFL = black soldier fly larvae; CO_2-eq = carbon dioxide equivalent; GHG = greenhouse gas; MWh = megawatt-hours; t = tons.

ANGOLA

TABLE 4A.38 Crop Waste, BSFL, and Frass, Angola

Crop	Total crop-related waste (tons)	BSFL produced (tons)		BSFL frass produced (tons)	
		10%	30%	10%	30%
Maize	2,248,611	224,861	674,583	224,861	674,583
Groundnut	652,082	65,208	195,625	65,208	195,625
Soybean	16,292	1,629	4,888	1,629	4,888
Sugarcane	259,408	25,941	77,822	25,941	77,822
Cassava	5,292,696	529,270	1,587,809	529,270	1,587,809
Total	**8,469,090**	**846,909**	**2,540,727**	**846,909**	**2,540,727**

Source: Original table for this publication.
Note: 10% = low conversion rate; 30% = high conversion rate; BSFL = black soldier fly larvae.

TABLE 4A.39 Livestock Protein Demand, Angola

Livestock/crop	Quantity	Total protein demand (tons)		
Chickens	36,500,000	188,355		
Pigs	4,512,098	123,271		
Goats	2,727,646	314,307		
Aquaculture (tons)	1,339	557		
Total		**626,489**		
BSFL dry matter from	**Total BSFL dry matter (tons)**	**Total crude protein (tons)**	**Total BSFL dry matter (tons)**	**Total crude protein (tons)**
	10% conversion rate		**30% conversion rate**	
Maize	76,453	30,581	229,358	91,743
Groundnut	22,171	8,868	66,512	26,605
Soybean	554	222	1,662	665
Sugarcane	8,820	3,528	26,460	10,584
Cassava	179,952	71,981	539,855	215,942
TOTAL	**287,949**	**115,180**	**863,847**	**345,539**
BSFL protein surplus (deficit)		**(511,309)**		**(280,950)**
Crude protein/dry matter for BSFL (%)	*40*			
Total protein demand provided by BSFL (%)		**18.4**		**55.2**

Source: Original table for this publication.
Note: BSFL = black soldier fly larvae.

TABLE 4A.40 Employment Generation, Angola

Direct employment		Indirect employment		Total employment	
10%	30%	10%	30%	10%	30%
57,590	172,769	158,948	476,844	216,538	649,613

Source: Original table for this publication.
Note: 10% = low conversion rate; 30% = high conversion rate.

TABLE 4A.41 GHG Emissions Reduction and Energy Savings Realized through BSFL-Derived Meal and Frass, Angola

Reduction/savings	10%	30%
GHG emissions reduction from BSFL meal (t CO_2-eq)	609,159	1,827,476
GHG emissions reduction from frass (t CO_2-eq)	643,651	1,930,952
Total GHG emissions reduction (t CO_2-eq)	1,252,810	3,758,429
• GHG emissions equivalent (passenger vehicles per year)	272,350	817,050
Energy savings from BSFL meal (MWh)	1,153,844	3,461,531
Energy savings from frass (MWh)	1,785,566	5,356,698
Total energy savings (MWh)	2,939,410	8,818,229
• Number of average homes in the northwestern United States that can be powered for one year by energy savings	268,439	805,318

Source: Original table for this publication.
Note: 10% = low conversion rate; 30% = high conversion rate; BSFL = black soldier fly larvae; CO_2-eq = carbon dioxide equivalent; GHG = greenhouse gas; MWh = megawatt-hours; t = tons.

NOTES

1. Retailers include bars, restaurants, and street vendors.
2. Packers and distributors distribute different agricultural products to local and regional markets.
3. The scale of the farms was self-reported as there is no current definition of small-, medium-, and large-scale insect farming systems.
4. https://patentscope.wipo.int/ (accessed December 2019).
5. The data are from FAOSTAT, the statistical database of the FAO.
6. Wastes and losses are referred to hereafter simply as wastes.
7. J. K. Tomberlin, personal communication.
8. Total crude protein available equals the dry meal mass multiplied by 40 percent.
9. All tons referred to in this study are metric (1,000 kg) unless otherwise stated.
10. A study from Michigan State University asserts that cattle (steers) can consume up to 20 percent stover on a dry matter basis in a feedlot operation without significantly affecting performance (Jean, Gould, and Anderson 2017).
11. Postharvest maize loss in Sub-Saharan Africa ranges from 5.6 to 25.5 percent (Affognon et al. 2015). The midpoint, 15.5 percent, is used for the estimation in this analysis.
12. Rounding errors result in 333 kg/ha rather than 334 kg/ha.
13. The research team referred to several sources, and the range of findings was from 40 percent (most sources reporting) to 48 percent. Results in laboratories have yielded as much as 60 percent. This study uses 40 percent, as recommended by an industry expert (J. K. Tomberlin).
14. The low 6 percent (38,834 tons) estimate and the high 17 percent (116,502 tons) estimate result from applying the conversion rates of 10 and 30 percent, respectively. This will be true for all subsequent low and high estimates in this analysis unless otherwise specified.

15. Price on September 25, 2019, on https://www.indexmundi.com/commodities/?commodity=soybean-meal.
16. For example, NPK 10-20-10 indicates a fertilizer with 10 percent nitrogen, 20 percent phosphorus, and 10 percent potassium.
17. Examples of BSFL frass advertised online include KIS Organics (NPK 3-2-4; https://www.kisorganics.com/products/natural-insect-fertilizer-frass) and The Critter Depot (NPK 5-3-2; https://www.thecritterdepot.com/blogs/news/nutritional-benefits-of-black-soldier-fly-larva-frass-critter-depot).
18. The frass derived from BSFL as reported in the tables in this report reflects the quantity of frass produced by BSFL. Decomposed substrate and chitin may be mixed in with the frass, thus comprising a biofertilizer that is mostly, but not strictly, frass.
19. Price on September 25, 2019, based on https://articles2.marketrealist.com/2016/02/update-npk-fertilizer-price-trends/.
20. Price on March 8, 2020, based on https://buildasoil.com/products/premium-insect-frass?variant=45848174802.
21. Direct jobs are specific to an industry, whereas indirect jobs are created outside the specific industry, that is, jobs in supporting industries. As examples, direct jobs associated with BSF include workers hired for breeding and processing, whereas indirect jobs include transportation services used for transporting frass and dry meal. For a detailed explanation of employment multipliers, refer to Bevins (2019).
22. Type I employment multipliers and effects by SU114 industry and sector (market, government, and nonprofit institutions serving households); reference year 2010.
23. Given frass production of 285,545 tons (at 10 percent conversion) to 856,635 tons (at 30 percent conversion).

REFERENCES

Affognon, H., C. Mutungi, P. Sanginga, and C. Borgemeister. 2015. "Unpacking Post-Harvest Losses in Sub-Saharan Africa: A Meta-Analysis." *World Development* 66: 49–68.

BAKERpedia. 2020. "Extraction Rate." BAKERpedia. https://bakerpedia.com/processes/extraction-rate/.

Berazneva, J. 2013. "Economic Value of Crop Residues in African Smallholder Agriculture." https://ageconsearch.umn.edu/record/150367/files/Berazneva-AAEA2013-final.pdf.

Bevins, J. 2019. "Updated Employment Multipliers for the U.S. Economy." Economic Policy Institute, Washington, DC. https://www.epi.org/publication/updated-employment-multipliers-for-the-u-s-economy/.

Brentrup, F., A. Hoxha, and B. Christensen. 2016. "Carbon Footprint Analysis of Mineral Fertilizer Production in Europe and Other World Regions." Tenth International Conference on Life Cycle Assessment of Food, University College Dublin, Ireland.

Diener, S., S. Semiyaga, C. B. Niwagaba, A. M. Muspratt, J. B. Gning, M. Mbéguéré, et al. 2014. "A Value Proposition: Resource Recovery from Faecal Sludge: Can It Be the Driver for Improved Sanitation?" *Resources, Conservation, and Recycling* 88: 32–38.

Dobermann, D., L. Michaelson, and L. M. Field. 2019. "The Effect of an Initial High-Quality Feeding Regime on the Survival of *Gryllus bimaculatus* (Black Cricket) on Bio-Waste." *Journal of Insects as Food and Feed* 5: 117–23.

Drew, D. J. W., and E. Pieterse. 2015. "Markets, Money and Maggots." *Journal of Insects as Food and Feed* 1 (3): 227–31.

EFSA (European Food Safety Authority). 2015. "Risk Profile Related to Production and Consumption of Insects as Food and Feed." *EFSA Journal* 13: 4257.

EPA (United States Environmental Protection Agency). 2018. "Greenhouse Gas Emissions from a Typical Passenger Vehicle." EPA, Washington, DC. https://www.epa.gov/greenvehicles/greenhouse-gas-emissions-typical-passenger-vehicle.

FAO (Food and Agriculture Organization of the United Nations). 2011. "Global Food Losses and Waste." FAO, Rome. http://www.fao.org/3/a-i2697e.pdf.

Food Price Monitoring Committee. 2003. *Food Price Trends*, vol. 4, chapter 2. Food Price Monitoring Committee, Government of the Republic of South Africa. https://www.google.com/url?sa=t&rct=j&q=&esrc=s&source=web&cd=1&ved=2ahUKEwjnoOynvJHkAhUkSN8KHWgmC6cQFjAAegQIARAC&url=https%3A%2F%2Fwww.nda.agric.za%2Fdocs%2FGenReports%2FFPMC%2FVol4_Chap2.pdf&usg=AOvVaw0CYti-4vz7JNW0CZH73lmo.

Ganda, H., E. T. Zannou-Boukari, M. Kenis, C. A. A. M. Chrysostome, and G. A. Mensah. 2019. "Potentials of Animal, Crop and Agri-Food Wastes for the Production of Fly Larvae." *Journal of Insects as Food and Feed* 5: 59–67.

Halloran, A. 2017. "The Impact of Cricket Farming on Rural Livelihoods, Nutrition and the Environment in Thailand and Kenya." PhD thesis, Department of Nutrition, Exercise and Sports, University of Copenhagen, Denmark. doi:10.13140/RG.2.2.23820.00641.

Halloran, A., and R. Flore. 2018. "A New World of Ingredients: Aspiring Chefs' Opinions on Insects in Gastronomy." In *Edible Insects in Sustainable Food Systems*, edited by A. Halloran, R. Flore, P. Vantomme, and N. Roos, 129–37. Springer International Publishing. doi:10.1007/978-3-319-74011-9_8.

Halloran, A., Y. Hanboonsong, N. Roos, and S. Bruun. 2017. "Life Cycle Assessment of Cricket Farming in North-Eastern Thailand." *Journal of Cleaner Production* 156: 83–94.

Heuzé, V., G. Tran, and S. Kaushik. 2020. "Soybean Meal." Feedipedia, a Programme by INRA, CIRAD, AFZ and FAO (March 4, 2020). https://feedipedia.org/node/674.

Homann, A. M. M., M. A. A. Ayieko, S. O. O. Konyole, and N. Roos. 2017. "Acceptability of Biscuits Containing 10% Cricket (Acheta domesticus) Compared to Milk Biscuits among 5-10-Year-Old Kenyan Schoolchildren." *Journal of Insects as Food and Feed* 3: 95–103.

IEA (International Energy Agency). 2014. *2014 Africa Energy Outlook*. Paris: IEA.

IITA (International Institute of Tropical Agriculture). 2020. "Maize." IITA, Ibadan, Nigeria. https://www.iita.org/cropsnew/maize/.

IPIFF (International Platform of Insects for Food and Feed). 2019. "The European Insect Sector Today: Challenges, Opportunities and Regulatory Landscape." IPIFF, Brussels, Belgium. http://ipiff.org/wpcontent/uploads/2019/12/2019IPIFF_VisionPaper_updated.pdf.

Jean, M., K. Gould, and E. Anderson. 2017. "The Michigan Corn Stover Project—Part 3: Cattle Feeding Study." Michigan State University, MSU Extension, East Lansing. https://www.canr.msu.edu/news/the_michigan_corn_stover_project_part_3_cattle_feeding_study.

Kelemu, S., S. Niassy, B. Torto, K. Fiaboe, H. Affognon, H. Tonnang, et al. 2015. "African Edible Insects for Food and Feed: Inventory, Diversity, Commonalities and Contribution to Food Security." *Journal of Insects as Food and Feed* 1 (2): 103–19.

Kenis, M., B. Bouwassi, H. Boafo, E. Devic, R. Han, G. Koko, et al. 2018. "Small-Scale Fly Larvae Production for Animal Feed." In *Edible Insects in Sustainable Food Systems*, edited by A. Halloran, R. Flore, P. Vantomme, and N. Roos. Springer International Publishing. doi:10.1007/978-3-319-74011-9_15.

Kinyuru, J. N., and C. Kipkoech. 2018. "Production and Growth Parameters of Edible Crickets: Experiences from a Farm in a High Altitude, Cooler Region of Kenya." *Journal of Insects as Food and Feed* 4: 247–51.

Kinyuru, J. N., and N. W. Ndung'u. 2020. "Promoting Edible Insects in Kenya: Historical, Present and Future Perspectives towards Establishment of a Sustainable Value Chain." *Journal of Insects as Food and Feed* 6 (1): 51–58.

Koné, N., M. Sylla, S. Nacambo, and M. Kenis. 2017. "Production of House Fly Larvae for Animal Feed through Natural Oviposition." *Journal of Insects as Food and Feed* 3: 177–86.

Lalander, C., S. Diener, M. E. Magri, C. Zurbrügg, A. Lindström, and B. Vinnerås. 2013. "Faecal Sludge Management with the Larvae of the Black Soldier Fly (Hermetia illucens): From a Hygiene Aspect." *Science of the Total Environment* 458–60: 312–318.

Li, Q., L. Zheng, N. Qiu, H. Cai, J. K. Tomberlin, and Z. Yu. 2011. "Bioconversion of Dairy Manure by Black Soldier Fly (Diptera: Stratiomyidae) for Biodiesel and Sugar Production." *Waste Management* 31: 1316–20.

Lynley, M. 2018. "Ovipost Wants to Help Drop the Labor Cost of Building Cricket Farms." Techcrunch, March 8, 2018. https://techcrunch.com/2018/03/08/ovipost-wants-to-help-drop-the-labor-cost-of-growing-cricket-farms/.

Macombe, C., S. Le Feon, J. Aubin, and F. Maillard. 2019. "Marketing and Social Effects of Industrial Scale Insect Value Chains in Europe: Case of Mealworm for Feed in France." *Journal of Insects as Food and Feed* 5: 215–24.

MAFRA (Ministry of Agriculture, Food and Rural Affairs). 2019. "Survey of the Current Status of Korean Insect Industry." MAFRA, Sejong City, Republic of Korea.

Magara, H. J. O., C. M. Tanga, M. A. Ayieko, S. Hugel, S. A. Mohamed, F. M. Khamis, et al. 2019. "Performance of Newly Described Native Edible Cricket Scapsipedus icipe (Orthoptera: Gryllidae) on Various Diets of Relevance for Farming." *Journal of Economic Entomology* 112 (2): 653–64.

Makkar, H. P. S., G. Tran, V. Heuzé, and P. Ankers. 2014. "State-of-the-Art on Use of Insects as Animal Feed." *Animal Feed Science and Technology* 197: 1–33. https://www.sciencedirect.com/science/article/abs/pii/S0377840114002326.

Manditsera, F. A., C. M. M. Lakemond, V. Fogliano, C. J. Zvidzai, and P. A. Luning. 2018. "Consumption Patterns of Edible Insects in Rural and Urban Areas of Zimbabwe: Taste, Nutritional Value and Availability Are Key Elements for Keeping the Insect Eating Habit." *Food Security* 10: 561–70.

Martínez-Sánchez, A., C. Magaña, M. Saloña, and S. Rojo. 2011. "First Record of Hermetia illucens (Diptera: Stratiomyidae) on Human Corpses in Iberian Peninsula." *Forensic Science International* 206 (1-3): e76–e78.

Miech, P., O. Berggren, J. E. Lindberg, T. Chhay, B. Khieu, and A. Jansson. 2016. "Growth and Survival of Reared Cambodian Field Crickets (Teleogryllus testaceus) Fed Weeds, Agricultural and Food Industry By-Products." *Journal of Insects as Food and Feed* 2: 285–92.

Morales-Ramos, J. A., M. G. Rojas, and A. T. Dossey. 2018. "Age-Dependent Food Utilisation of *Acheta domesticus* (Orthoptera: Gryllidae) in Small Groups at Two Temperatures." *Journal of Insects as Food and Feed* 4: 51–60.

Muafor, F. J., A. A. Gnetegha, P. Le Gall, and P. Levang. 2015. "Exploitation, Trade and Farming of Palm Weevil Grubs in Cameroon." Center for International Forestry Research, Bogor, Indonesia.

Neville, P. F., and T. D. Luckey. 1962. "Carbohydrate and Roughage Requirement of the Cricket, Acheta domesticus." *Nutrition* 78: 139–46.

Nikodinovic-Runic, J., M. Guzik, S. T. Kenny, R. Babu, A. Werker, and K. E. O'Connor. 2013. "Carbon-Rich Wastes as Feedstocks for Biodegradable Polymer (Polyhydroxyalkanoate) Production Using Bacteria." In *Advances in Applied Microbiology*, vol. 84, edited by S. Sariaslani and G. M. Gadd, 139–200. Elsevier.

Nyakeri, E. M. M., H. J. J. Ogola, M. A. A. Ayieko, and F. A. A. Amimo. 2016. "An Open System for Farming Black Soldier Fly Larvae as a Source of Proteins for Smallscale Poultry and Fish Production." *Journal of Insects as Food and Feed* 3: 51–56.

Orinda, M. A., R. O. Mosi, M. A. Ayieko, and F. A. Amimo. 2017. "Effects of Housing on Growth Performance of Common House Cricket (Acheta domesticus) and Field Cricket (Gryllus bimaculatus)." *Journal of Entomology and Zoology Studies* 5: 1664–68.

Parodi, A., I. J. M. De Boer, W. J. J. Gerrits, J. J. A. Van Loon, M. J. W. Heetkamp, J. Van Schelt, J. E. Bolhuis, et al. 2020. "Bioconversion Efficiencies, Greenhouse Gas and Ammonia Emissions during Black Soldier Fly Rearing: A Mass Balance Approach." *Journal of Cleaner Production* 271: 122488.

Persistence Market Research. 2019. *Edible Insects for Animal Feed Market: Global Industry Analysis*. New York: Persistence Market Research.

Peters, Marian. Undated. "2x More & 2x Less." Presentation, New Generation Nutrition B.V., Netherlands. https://europa.eu/capacity4dev/file/18202/download?token=cmTwwkFN.

Pomalègni, S. C. B., D. S. J. C. Gbemavo, C. P. Kpadé, M. Kenis, and G. A. Mensah. 2017. "Traditional Use of Fly Larvae by Small Poultry Farmers in Benin." *Journal of Insects as Food and Feed* 3: 187–92.

Purohit, C., and P. Rajyalakshmi. 2011. "Quality of Products Containing Defatted Groundnut Cake Flour." *Journal of Food Science and Technology* 48 (1): 26–35. https://www.ncbi.nlm.nih.gov/pmc/articles/PMC3551082/.

Rees, J., C. Wortmann, M. Drewnoski, K. Glewen, R. Pryor, and T. Whitney. 2018. "What Is the Value of Soybean Residue?" University of Nebraska-Lincoln. https://cropwatch.unl.edu/2018/what-value-soybean-residue.

Sanou, A. G., F. Sankara, S. Pousga, K. Coulibaly, J. P. Nacoulma, M. Kenis, et al. 2018. "Indigenous Practices in Poultry Farming Using Maggots in Western Burkina Faso." *Journal of Insects as Food and Feed* 4: 219–28.

Selenius, O. V. O., J. Korpela, S. Salminen, and C. G. Gallego. 2018. "Effect of Chitin and Chitooligosaccharide on In Vitro Growth of Lactobacillus rhamnosus GG and Escherichia coli TG." *Applied Food Biotechnology* 5: 163–72.

Sheppard, C. 1983. "House Fly and Lesser Fly Control Utilizing the Black Soldier Fly in Manure Management Systems for Caged Laying Hens." *Environmental Entomology* 12: 1439–42.

Sheppard, D. C., J. K. Tomberlin, J. A. Joyce, B. C. Kiser, and S. M. Sumner. 2002. "Rearing Methods for the Black Soldier Fly (Diptera: Stratiomyidae)." *Journal of Medical Entomology* 39: 695–98.

Shumo, M., I. M. Osuga, F. M. Khamis, C. M. Tanga, K. K. M. Fiaboe, S. Subramanian, et al. 2019. "The Nutritive Value of Black Soldier Fly Larvae Reared on Common Organic Waste Streams in Kenya." *Scientific Reports* 9, article 10110.

Sims, R., A. Flammini, M. Puri, and S. Bracco. 2015. *Opportunities for Agri-Food Chains to Become Energy-Smart*. Rome: Food and Agriculture Organization of the United Nations.

Sipponen, M. H., O. E. Mokinen, K. Rommi, R. L. Heinio, U. Holopainen-Mantila, S. Hokkanen, et al. 2018. "Biochemical and Sensory Characteristics of the Cricket and Mealworm Fractions from Supercritical Carbon Dioxide Extraction and Air Classification." *European Food Research and Technology* 244: 19–29.

Statista. 2020. "Sugar Cane Production Worldwide from 1965-2018." Statista, Hamburg, Germany. https://www.statista.com/statistics/249604/sugar-cane-production-worldwide/.

Stoops, J., D. Vandeweyer, S. Crauwels, C. Verreth, H. Boeckx, M. Van Der Borght, et al. 2017. "Minced Meat-Like Products from Mealworm Larvae (Tenebrio molitor and Alphitobius diaperinus): Microbial Dynamics during Production and Storage." *Innovative Food Science & Emerging Technologies* 41: 1–9.

Stull, V. J., M. Wamulume, M. I. Mwalukanga, A. Banda, R. S. Bergmans, and M. M. Bell. 2018. "'We Like Insects Here': Entomophagy and Society in a Zambian Village." *Agriculture and Human Values* 35: 867–83. https://doi.org/10.1007/s10460-018-9878-0.

Teravaninthorn, S., and G. Raballand. 2009. "Transport Prices and Costs in Africa: A Review of the Main International Corridors." World Bank, Washington, DC.

Tonneijck-Srpová, L., E. Venturini, K. N. Humblet-Hua, and M. E. Bruins. 2019. "Impact of Processing on Enzymatic Browning and Texturization of Yellow Mealworms." *Journal of Insects as Food and Feed* 5 (4): 267–77.

van Huis, A. 2019. "Manure and Flies: Biodegradation and/or Bioconversion?" *Journal of Insects as Food and Feed* 5: 55–58.

van Huis, A., D. G. A. B. Oonincx, S. Rojo, and J. K. Tomberlin. 2020. "Insects as Feed: House Fly or Black Soldier fly?" *Journal of Insects as Food and Feed* 6: 221–29.

Veenenbos, M. E., and D. G. A. B. Oonincx. 2017. "Carrot Supplementation Does Not Affect House Cricket Performance (*Acheta domesticus*)." *Journal of Insects as Food and Feed* 3: 217–21.

Vernooij, A. G., and T. Veldkamp. 2019. *Insects for Africa: Developing Business Opportunities for Insects in Animal Feed in Eastern Africa.* Wageningen, Netherlands: Wageningen Research.

WageIndicator Foundation. 2020. "Living Wage Series – Zimbabwe – January 2018." WageIndicator Foundation, Amsterdam, Netherlands. https://wageindicator.org/salary/living-wage/zimbabwe-living-wage-series-january-2018.

Wang, Y.-S., and M. Shelomi. 2017. "Review of Black Soldier Fly (Hermetia illucens) as Animal Feed and Human Food." *Foods* 6 (10): 91.

Washington State University. 2016. *Fertilizing with Manure and Other Organic Amendments.* Pullman, WA: Washington State University. https://pubs.extension.wsu.edu/fertilizing-with-manure.

Zafar, S. 2020. "Sugarcane Trash as Biomass Resource." *BioEnergy Consult*, September 27, 2020. https://www.bioenergyconsult.com/sugarcane-trash-biomass/.

CHAPTER FIVE

Understanding Hydroponics

HIGHLIGHTS

- Hydroponics is a climate-smart technology that grows nutritious food and can contribute to food security, job and livelihood creation, and environmental protection.
- The market for hydroponics was worth about US$8.1 billion in 2019 and will be worth US$16 billion by 2025 (MarketsandMarkets 2019)—a 12.1 percent annual growth rate.
- There are a variety of hydroponic systems, from simple to sophisticated and from open to closed. These systems include wick, deep water culture, ebb and flow, drip method, nutrient film technique (NFT), aquaponics, and aeroponics.
- Hydroponics requires several inputs. These include seeds, seedlings, labor, electricity, technical knowledge, building materials, and start-up capital and operating costs. Hydroponics also requires basic inputs such as water, nutrient solution, and a growing medium.
- Simple hydroponic systems do not require much labor. People without a formal education can rapidly acquire the skills needed to operate hydroponic systems.
- Hydroponic farming is possible in densely populated urban zones or water-scarce environments because it requires up to 75 percent less space than conventional farming methods.
- Hydroponics requires approximately 80 to 99 percent less water than traditional agriculture.

- Hydroponic plants have similar or higher amounts of nutrients than conventionally grown produce.
- Hydroponic crops have greater yields and require fewer pesticides than traditional agricultural crops.
- Hydroponic systems can hedge against climate change risks because they operate in climate-controlled conditions and are not exposed to temperature variations.
- Hydroponic systems can be profitable despite high start-up costs. In The Gambia, hydroponic production costs are US$2.30 for 1 square meter (m^2) of lettuce and US$3 for 1 m^2 of sweet pepper, whereas profits reach US$6 for lettuce and US$15 for sweet pepper.
- As hydroponic systems scale up, production costs decrease and profits come sooner. And processing hydroponic crops can improve their profitability.

This chapter examines the potential of hydroponics as a frontier agricultural technology within a circular food economy. Hydroponics is the process of growing plants in nutrient solutions instead of soil (Verner et al. 2017; Jensen 1997). Hydroponics complements insect farming, which produces animal source foods in a circular food production model, by producing nutritious fruits, vegetables, and grains. Hydroponics is a climate-smart technology that can contribute to food security, job and livelihood creation, and environmental protection. The first section in the chapter defines hydroponics and its history, and the second section describes the various types of hydroponic systems. The third section describes the inputs needed to operate a hydroponic system, principally the water, nutrient solution, and growing medium. The fourth section looks at the outputs and products produced by hydroponic systems, including nutritious produce for human consumption and animal feed. The fifth section then examines the benefits from using hydroponics instead of traditional soil agriculture. These benefits include greater yields, high-value products, reduced land and water use, energy efficiency, pest management, and specific benefits for people in countries affected by fragility, conflict, and violence (FCV). The sixth section presents some of the limitations of hydroponic systems and how to mitigate them. These include capital requirements, running costs, and limited knowledge on the industry and its operations.

ABOUT HYDROPONICS

Hydroponics is an expanding practice in Africa that can grow crops quickly without soil. Growing hydroponic crops is not new; however, using hydroponics to achieve development goals, especially in harsh climates, is an innovative approach to development. Growing plants in nutrient-rich water has been practiced for centuries. Early examples of hydroponic growing include Babylon's hanging gardens and the Aztecs' floating gardens in Mexico. In 1929, a professor from the University of California, Berkeley began growing plants in a soilless medium and coined the term "hydroponics." In traditional farming, soil stores the various nutrients required for plant growth. When water saturates the soil, the water picks up these nutrients and is absorbed by the plant's roots (Campbell and Reece 2002), moving to the plant's shoots, leaves, and fruits. In hydroponics, the need for soil is removed by feeding nutrient-rich water directly to the plant. As this chapter shows, hydroponic farming is increasingly becoming an important crop production technology in FCV contexts, including refugee camps and arid host communities in East and West Africa. For example, hydroponic systems have produced animal feed in Chad and human food in Kenya, Sudan, and Zambia. In Zambia, the World Food Programme (WFP) provided smallholder farmers hydroponic greenhouses to increase their resilience to climate variability. Each greenhouse holds approximately 2,000 plants that produce 1,300 kilograms (kg) of vegetables per month. The global hydroponic market was estimated at US$8.1 billion in 2019 and projected to grow at an annual growth rate of 12.1 percent to reach US$16 billion by 2025 (MarketsandMarkets 2019).

Hydroponics can be a livelihood option for displaced populations and other vulnerable populations. Most refugees do not expect to be displaced for long, but in reality displacement lasts 10 years on average (Devictor and Do 2017). As a result, there is a need to complement short-term humanitarian assistance to displaced communities with longer-term development efforts. Displaced populations generally have an entrepreneurial spirit and motivation to work. Moreover, many displaced populations worked in agriculture before migrating and could apply these skills to their new communities but often lack access to local land, knowledge, and resources. Providing this access is where the development community can make a difference. Generally, investing in agriculture is an effective long-term strategy to create jobs, build livelihoods, and strengthen food security (Verner 2016). Meanwhile, investing in hydroponics can be particularly beneficial for displaced persons, who often relocate to areas that are less suitable for conventional agriculture, such as arid resettlement camps or densely populated urban spaces. Providing jobs, livelihoods, and greater food security for the displaced will also reduce the resource burden on host communities. In the refugee and host community context, hydroponics has several advantages over conventional farming methods, including year-round production, increased yields, faster growth rates, access to fresh vegetables, greater potential for self-sustainability, and important social and psychological benefits for displaced individuals (WFP Kenya 2020). Moreover, simple hydroponic systems can be constructed with locally available and recycled materials, such as jerrycans or other discarded receptacles (Verner et al. 2017). Box 5.1 describes how the WFP used hydroponics to achieve development goals in a Kenyan refugee camp.

The hydroponic market was worth US$9.5 billion in 2020, with private sector investment driving its 11.3 percent compound annual growth rate (Intrado 2021). Investments in the hydroponic farming industry have grown rapidly over the past half-decade. From 2016 to 2017, venture capital

BOX 5.1 Hydroponic Pilot Project in Kenya's Kakuma Refugee Camp

In December 2018, the World Food Programme launched a pilot project for simplified hydroponic crop production in Kenya's Kakuma refugee camp. The pilot's immediate objective was to determine the viability of hydroponics as a sustainable production method for refugee and host communities located in dry, harsh climates that are not suitable for agriculture. The pilot's medium- to long-term objective was to improve refugees' livelihoods and income-generating opportunities. Nine different hydroponic systems, including horizontal and vertical systems, were tested at 38 sites at the household and community levels. By the time the pilot project closed in December 2019, it had completed seven harvest cycles with 74 percent of the units remaining active and productive throughout the project's duration. In total, the pilot helped 1,500 beneficiaries produce hydroponically grown vegetables (WFP Kenya 2020). Nearly 50 percent of the beneficiaries were women and 43 percent were youth, including 100 youth affected by scurvy, which was caused by prolonged vitamin C deficiency.

The pilot showed that hydroponics uses less water, has higher crop yields, and has faster growth rates than conventional field agriculture. Hydroponics used between 82 and 92 percent less water than conventional farming for growing kale, spinach, and cowpeas. One circular garden and two hydro crates, which together require 4 square meters of land, produced 2,780 grams of spinach, compared with 188 grams of spinach produced on a conventional farm using the same amount of space. That is 15 times more produce from the hydroponic system than from conventional farming. Meanwhile, one hanging garden, which accommodates only 64 plants, produced 446 grams of spinach, whereas the conventionally farmed plot, which accommodates 89 plants, produced only 200 grams of spinach. For cowpeas, the hanging garden required more time than conventional farming to reach harvest, 44 days compared with 40 days. The pilot also showed that the production costs for hydroponic systems decrease and profits come sooner as production units are scaled up. These results are mentioned throughout this chapter.

funding for vertical farming increased from US$36 million to US$271 million, a 753 percent increase. Much of this funding was directed to large companies with strong tech initiatives, such as Plenty USA and others. Additionally, crowdfunding on Kickstarter and Indiegogo for a variety of indoor agricultural initiatives increased by 1,000 percent, from US$2.8 million to US$28 million (Clark 2018). In Dubai, the Abu Dhabi Investment Office invested in Aerofarms USA to build a hydroponic farm. Barclays UK invested in 80 Acres, another hydroponic farm. In 2019, a group of investors invested US$100 million in InFarm,[1] an indoor agricultural supplier, to scale the company's growth in Europe and the United States (Agritecture 2019). Overall, venture capital investments in the indoor agricultural industry, including hydroponics, reached a new high of US$565 million in 2020, which was a 50 percent increase from 2019 (iGrow 2020). The increasing success of the commercial hydroponic industry owes to the success of both system input companies and hydroponic crop producers. Leading system input companies include Signify Holdings

(Netherlands), Argus Control Systems (Canada), Heliospectra AB (Sweden), and Scotts Miracle Gro (United States). Leading hydroponic crop producers include AeroFarms (United States), Terra Tech Corp (United States), Triton Foodworks (India), and Emirates Hydroponic Farm (United Arab Emirates). The hydroponic industry is expected to be worth US$22.2 billion by 2028 (Intrado 2021).

TYPES OF HYDROPONIC SYSTEMS

There are a variety of hydroponic systems, from simple to sophisticated. There are numerous approaches to hydroponics—including aeroponics, fogponics, aquaponics, dryponics, and others. Some of these approaches are open and some are closed, but each follows the same principle: plants grow, without soil, in a circulating nutrient-rich water system (Wootton-Beard 2019). Among the different hydroponic approaches are various systems, including wick, deep water culture, ebb and flow, drip method, NFT, aquaponics, and aeroponics. These systems share many features but fundamentally differ in how they manage the nutrient solution. The most popular systems are deep water culture, drip method, and NFT (Resh 1995). The grower chooses the system depending on the type of plant and the facility's limitations, whether a lack of growing space or materials (Jensen 1997). Figure 5.1 describes the different systems and provides examples of how each is set up.

Hydroponic systems can generally be delineated into open and closed systems. Open systems, also known as "run-to-waste systems," do not reuse water. The nutrient solution flows through the system only once and is discarded (Jensen 1997; Nederhoff and Stanghellini 2010). Open systems provide two primary advantages over closed systems: (1) they do not require nutrient solution maintenance, and (2) they reduce the risk of acquiring infectious plant pathogens (Jones 2016). Despite these advantages, open systems are known to be wasteful of water and nutrients (Nederhoff and Stanghellini 2010), which may not be appropriate for arid, water-scarce regions. By contrast, closed systems recirculate the nutrient solution for an unspecified length of time (Lykas et al. 2006). These systems add water and nutrients as necessary instead of replacing the entire solution after each use (Jensen 1997; Nederhoff and Stanghellini 2010). The nutrient solution is regularly monitored and adjusted to maintain proper nutrient ratios. As a result, closed hydroponic systems use 20 to 40 percent less water and nutrients than open hydroponic systems, but closed systems require more monitoring and maintenance. This need arises because ions accumulate as the nutrient solution recirculates (Lykas et al. 2006). And recirculation requires reservoirs and pumping systems that must be monitored and maintained. This infrastructure can be susceptible to failure if it is not managed well (Nederhoff and Stanghellini 2010).

Wick techniques are open systems that comprise raised garden beds sitting above a water reservoir. This is the most common hydroponic technique and the easiest to set up and maintain (Wootton-Beard 2019). In areas with

FIGURE 5.1 Hydroponic Systems and How They Are Set Up

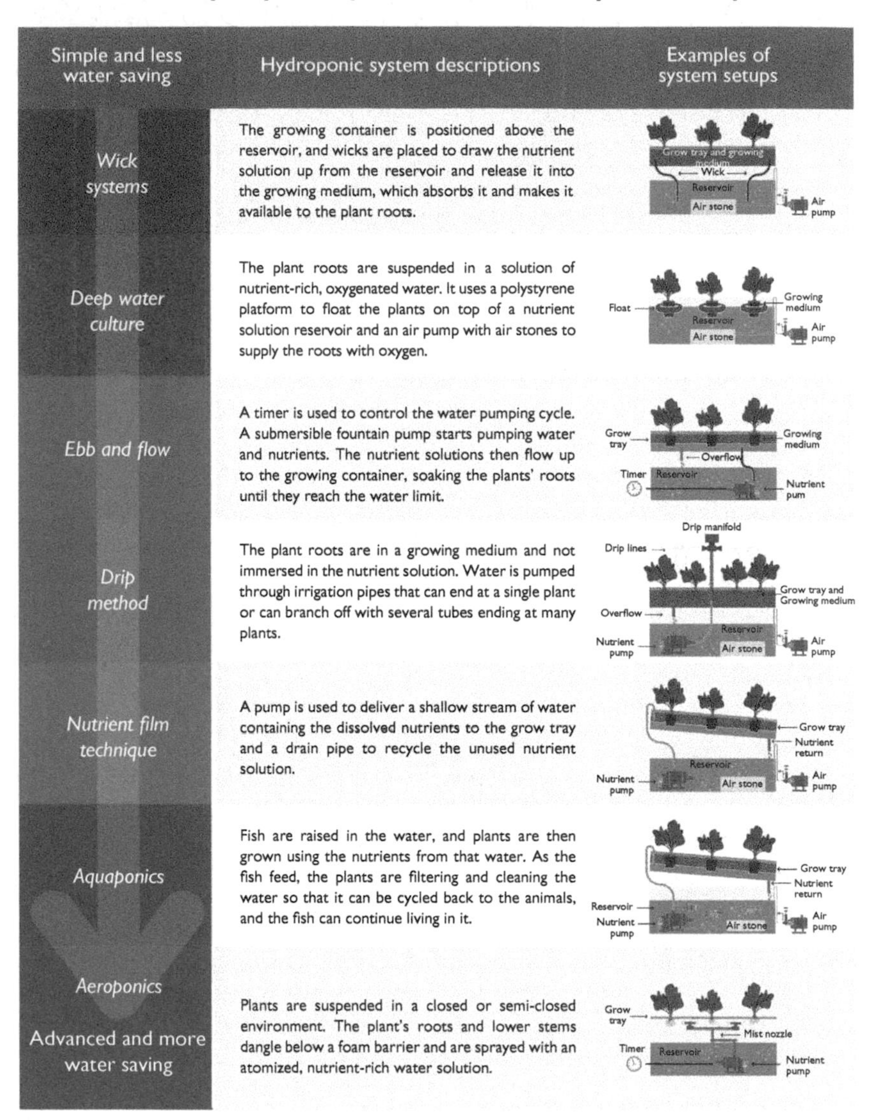

Source: Original figure for this publication.

nutrient-poor soils, growers embrace these techniques because they reduce the use of irrigation water by 50 percent.[2] Wick techniques consist of a small reservoir beneath a simple raised frame onto which bags of growing medium, or substrate, are placed. Water enters the water reservoir through a pipe and—by capillary action—is drawn upward through the wicks into the root zone of the plants in the bed above, enabling the plants to absorb the amount of water they need. In this technique, there is no need for overhead watering and less water is lost through evaporation. The roots growing in the substrate have a continuous supply of water, oxygen, and nutrients. The wick technique is low-cost and easy to maintain, especially for beginner growers since it does not require air or water pumps. Thus, despite being an open system, it works well when there are limited or unreliable technical inputs, such as intermittent electricity, and where troubleshooting and technical assistance are not readily available. For example, women in West Bank and Gaza use this system despite limited access to inputs and materials (photo 5.1).

Deep water culture is an inexpensive and open hydroponic technique that is easy to set up. The system uses a polystyrene platform to float the plants on top of a nutrient solution reservoir and an air pump with air stones to supply the roots with oxygen. Water culture systems are ideal for growing leafy greens, such as lettuce, because these plants grow rapidly and the systems can provide large amounts of water to propel that growth. The Kratky method, one form of deep water culture, demonstrates these attributes. In this method, the farmer uses a watertight container, such as a trash bin or five-gallon plastic storage container, filled with the nutrient solution (see photo 5.2). Plants are grown

PHOTO 5.1 Example of a Wick System with Multiple Wicking Beds in West Bank and Gaza

PHOTO 5.2 Kratky Bucket System

Photograph ©Eyal Barkan / FARM-IT. Used with permission from Eyal Barkan. Further permission required for reuse.

in net pots on top of the container's cover and continuously watered when the growing medium, or "soil," is moistened by capillary action. The plants' roots are partially submerged in the water and the other parts access oxygen, creating moist air around the plant. The Kratky method is inexpensive because it does not require any maintenance labor except for when planting, transplanting, or harvesting. This method does not require pumps or electricity, so the aeration and circulation costs that are common in other hydroponic systems are avoided (Kratky 2009).

The ebb and flow system, also called flood and drain, is a common hydroponic system that uses basic materials and is simple to set up. This method fills planted pots with inert growing medium and places the pots inside a tray or container. It then floods the plant roots with a nutrient solution and drains that solution back into the reservoir, which makes it a closed system and thus more water-saving than the previous methods described. The tray or container is automatically filled several times a day by a pump and timer during the growing season.

The drip system is another closed hydroponic technique that uses water-circulating drip emitters. The drip emitters drip water in every plant's pot of growing medium rather than spraying it into the air or running along the soil, which is the case in traditional drip irrigation. After the water passes through each plant's pot, it returns to the water reservoir and is recycled through the system again, making the system highly water efficient. Plants can be grown in pots, trays, and buckets. The system requires electricity to power a submersible pump that disperses the water and an air stone that mixes the water in the reservoir.

NFT circulates water though long plastic grow trays filled with plants. Small plastic net cups, filled with a growing medium, hold each plant. The grower sets each plant's water level depending on the plant's maturity. When the plants are younger, the water level is higher so the shallow roots can reach the water. Once the plants' roots mature, the water level is lower to promote root growth. The nutrient solution gets pumped through the system and circulates past the plant roots.

Aquaponics combines hydroponics with recirculating aquaculture to produce fish. Recirculation aquaculture is a technology for farming fish or other aquatic organisms by reusing the hydroponic water from plant production (Bregnballe 2015). The system cultivates plants in water while simultaneously raising fish in land-based tanks. Aquaponics recycles 95 to 99 percent of the system's water, distributing nutrients throughout the system. It generates the same type of produce as other hydroponic systems. Hydroponics and aquaponics generate similar produce yields; however, aquaponics also produces fish, which is an animal source food. Aquaponic systems typically use less water than the other hydroponic systems, with the exception of aeroponics. Since fish play a key role in the system and would die in water containing chemical pesticides, aquaponics grows chemical-free, all-natural produce. Figure 5.2 describes the aquaponics cycle.

Aeroponics is a relatively new method for growing edible plants and the most sophisticated hydroponic method. Aeroponics sprays enclosed plant roots with a fine mist of nutrient-laden water through a pressurized nozzle. Aeroponics requires 70 percent less water than other hydroponic systems, while delivering the same amount of nutrients to the roots. Recent advances in nozzle designs have improved the system's reliability by eliminating clogging from mineral deposits, which was a major issue in earlier aeroponic models. The system's improved reliability led to many vertical farms adopting aeroponics as their main growing method. Figure 5.3 lists the advantages and disadvantages of each type of hydroponic system.

Vertical farming is another form of space-saving agriculture that can utilize various hydroponic methods. Vertical farms grow produce in vertically stacked layers, which are vertically inclined or integrated in other tall structures, like towers or building walls, to increase plant growth (Christie 2014). Vertical farms are simply a way of saving space for agriculture in controlled environments. Vertical farming can employ different types of hydroponic systems. These can include simple NFT systems or sophisticated multistory aeroponic systems.

FIGURE 5.2 Aquaponics Cycle

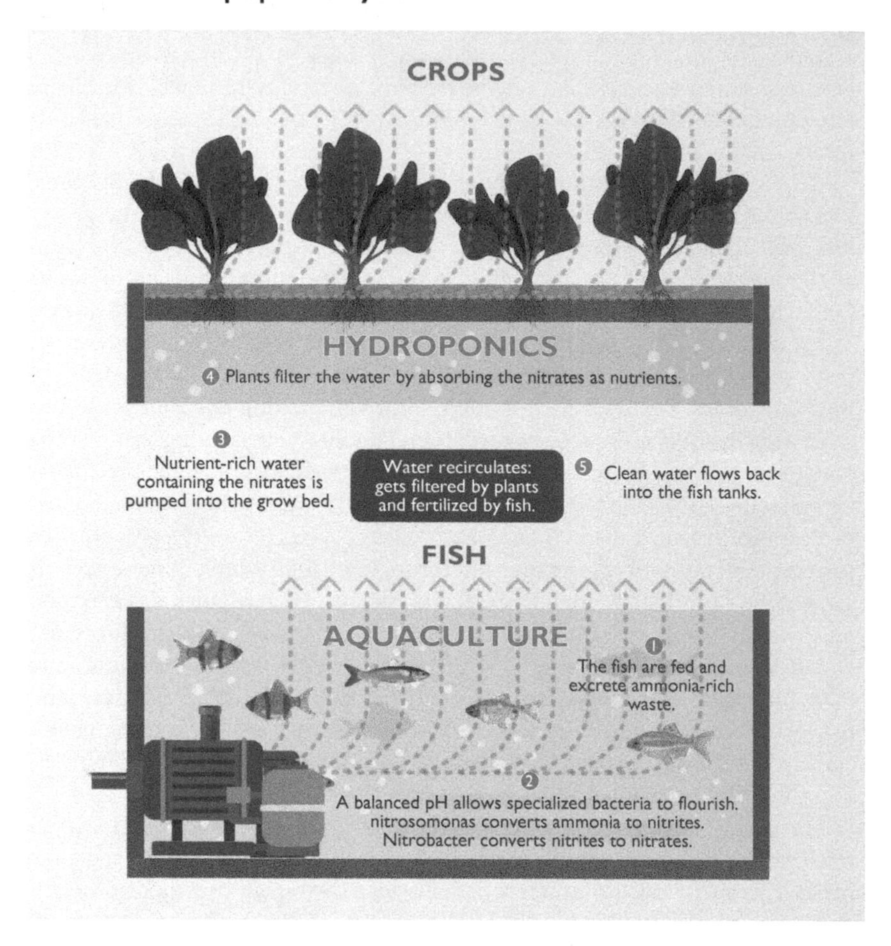

Source: Original figure for this publication.

For example, Aerofarms, which is located in the United States, uses multistory aeroponics to grow fruits, vegetables, and nonedible plants for drugs, vaccines, and biofuels. South Africa also has several commercial indoor vertical farming companies, and a few major cities in Africa have set up vertical farming facilities. Today, China; Japan; Taiwan, China; Singapore; the United States; and a few countries in Western Europe lead the world in vertical farming. Vertical farms can be constructed in small spaces such as urban rooftops and other unused spaces. They use less land than traditional agriculture while generating higher crop yields per unit area.

FIGURE 5.3 Advantages and Disadvantages of Hydroponic Systems

Type of system	Advantages	Disadvantages
Wick systems	• Affordable • Simple setup • Low maintenance • No nutrient pump or electricity needed	• Limited oxygen access • Slower growth rate • No nutrient recirculation • Prone to algae growth • Less efficient than other hydroponic methods • Salt buildup needs flushing
Deep water culture	• Inexpensive • Simple setup • Low maintenance • No nutrient pump or electricity needed with Kratky method • Reliable	• Risk of root rot if not cleaned regularly • Slower growth rate • Must top water until roots are long enough to fall into the nutrient solution • Must frequently refill reservoir
Ebb and flow	• Affordable • Low maintenance • Excess nutrient solution recirculates	• Prone to algae growth • Technical malfunctions could result in crop loss
Drip method	• Excess nutrient solution recirculates • Sufficient oxygen flow	• Prone to clogging • Prone to algae growth • Requires regular cleaning
Nutrient film technique	• Excess nutrient solution recirculation • Plentiful oxygen flow • Space sufficient	• Prone to clogging • Technical malfunctions could result in crop loss
Aquaponics	• Ability to raise fish • Recycles 95%–99% of water • Completely organic • Uses 90% less water than traditional farming • No chemical pesticides	• High start-up costs • High risk of system failure • Needs regular monitoring • High energy usage • Needs technical expertise • Needs reliable electricity
Aeroponics	• Maximum nutrient absorption • Excess nutrients recirculate • Plentiful oxygen flow • Space sufficient • Approximately 70% less water than hydroponics	• Prone to clogging • Technical malfunctions could result in crop loss • High-tech • Time intensive • Poorly suited to thick organic-based nutrients and additives

Source: Original figure for this publication.

REQUIRED INPUTS

Hydroponics requires several inputs and produces certain outcomes. Figure 5.4 shows that a hydroponic operation can have many inputs. These include seeds, seedlings, labor, nutrients, electricity, technical knowledge, building materials, community assessments, maintenance and troubleshooting, and start-up capital and operating costs. Hydroponics also requires basic inputs such as water, nutrient solution, and a growing medium. These inputs lead to several outcomes, such as highly nutritious food, increased incomes, and others, as listed in figure 5.4.

Nutrient Solution

The main input into any hydroponic system is the nutrient solution. This solution combines water and nutrient salts at specific concentrations to optimize a

FIGURE 5.4 Inputs into and Outcomes of Aquaponics and Hydroponics

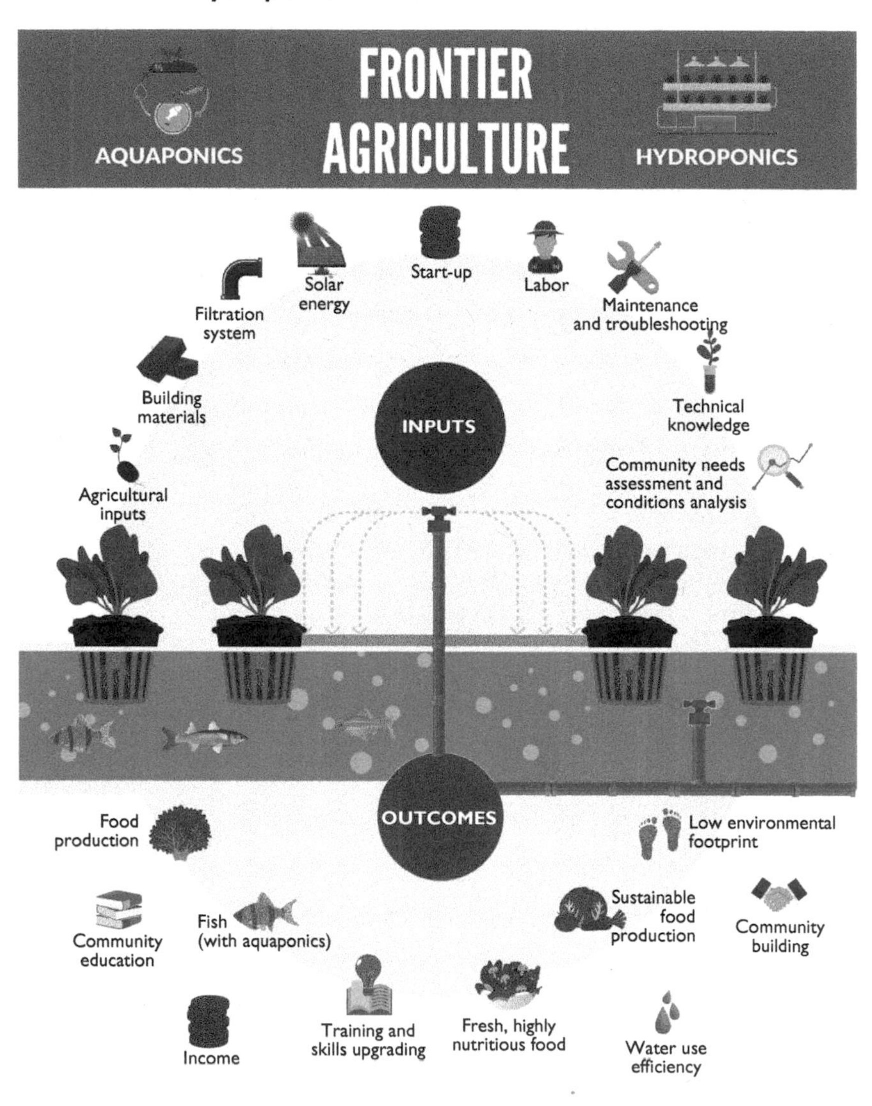

Source: Original figure for this publication.

plant's growth and health (Hoagland and Arnon 1950; Graves 1980; Jones 2016; Resh 2013). The aim of any hydroponic system is to deliver an optimized nutrient solution to plant roots (Wootton-Beard 2019). The interaction between the plant, growing medium, and nutrient solution determines the efficacy of the plant's growing environment. The nutrient solution's composition is controlled by the farmer and can be delivered to plants in the quantities and concentrations needed. This ability to control the solution allows hydroponics to optimize the nutrient intake of plants, producing high yields, while still minimizing overall water and nutrient usage. Scientists and horticulturalists understand which nutrients benefit plants and determine which to use based on their "essentiality." A nutrient is essential if its absence inhibits plant growth or reproduction (Arnon and Stout 1939). As such, certain plant cultivars thrive in hydroponics depending largely on the nutrient levels in the solution (Christie 2014). The three primary macronutrients are nitrogen, phosphorus, and potassium because they are the most commonly deficient in plants (Campbell and Reece 2002). The right combination of macronutrients and micronutrients affects the plant's flavor and nutritive value.

A nutrient solution's pH, which is a measure of acidity and alkalinity, is a common consideration used in hydroponic growing. The pH of a plant's root zone determines what nutrients are available to the plant, as plants can uptake certain key ions within a pH range of 5 and 7 only (Graves 1983). In traditional agriculture, the soil, not the plant, maintains the pH range in the root zone (Campbell and Reece 2002). Any change in pH will trigger a reaction in the plants such as changes in their health or growth patterns (Graves 1983; Jones 2016; Resh 2013; Christie 2014).

Water

Water is the primary ingredient in a nutrient solution, but water impurities can inhibit hydroponic systems. Most tap water contains a variety of ions (Spensley, Winsor, and Cooper 1978) or chlorine residuals from chemical treatments (Graves 1983). Therefore, using this water for hydroponics could contribute to toxic ion buildup in the nutrient solution (Lykas et al. 2006) or interfere with nutrient solution measurements (Resh 2013). Tap water usually has a pH near or above 7, which can adversely affect a plant's nutrient uptake. To combat these impurities requires some type of filtration system, such as a reverse osmosis unit (Resh 2013).

Hydroponic systems conserve more water than soil-based cultivation. This is especially beneficial in arid or climate change–affected environments. The water efficiency of hydroponics results from the ability to direct water to where it is most advantageous to plant growth (Wootton-Beard 2019). Plants consume the same amounts of water in hydroponics and conventional soil methods; however, hydroponic systems deliver the water more efficiently (Sanchez 2007). Closed hydroponic farming uses 80 to 99 percent less water than conventional irrigated farming since plants consume only the water they need while unused water recycles back into the reservoir for reuse. Box 5.2 (with figure B5.2.1) shows the more efficient water use in hydroponic systems

BOX 5.2 Comparing Lettuce Yields, Water Usage, and Growing Seasons between Traditional Soil Farming and Two Hydroponic Techniques—the Wicking Bed and Nutrient Film Techniques—in West Bank and Gaza

The Applied Research Institute in Jerusalem compared lettuce production in West Bank and Gaza using hydroponic systems and traditional soil methods. Figure B5.2.1 shows the annual lettuce yields, water usage, and number of growing seasons of traditional soil-based agriculture and two hydroponic systems: the wicking bed technique and the nutrient film technique (NFT). The data are based on 1 square meter of growing area for each of the three scenarios. The results show that the annual water requirement, 700 cubic meters (m^3), for the two growing seasons of the traditional soil plot was 500 percent higher than that of the four annual growing seasons associated with the wicking bed technique, at 118 m^3, and 700 percent higher than that of the four annual growing seasons for NFT, at 87 m^3. Water usage for NFT was 26.3 percent lower than for the wicking bed technique. The lettuce yield for soil farming was considerably lower than the yields of the hydroponic systems despite the substantially higher volume of water required for traditional soil farming. The traditional soil plot yielded 12 kg of lettuce, whereas the wicking bed technique resulted in 37 kg of lettuce (208 percent more than from traditional soil farming) and NFT yielded 45 kg (275 percent more than from soil farming and 21.6 percent more than from the wicking bed technique).[a]

FIGURE B5.2.1 Lettuce Yield, Water Use, and Number of Growing Seasons per Year for Two Hydroponic Systems and the Traditional Soil Method in West Bank and Gaza

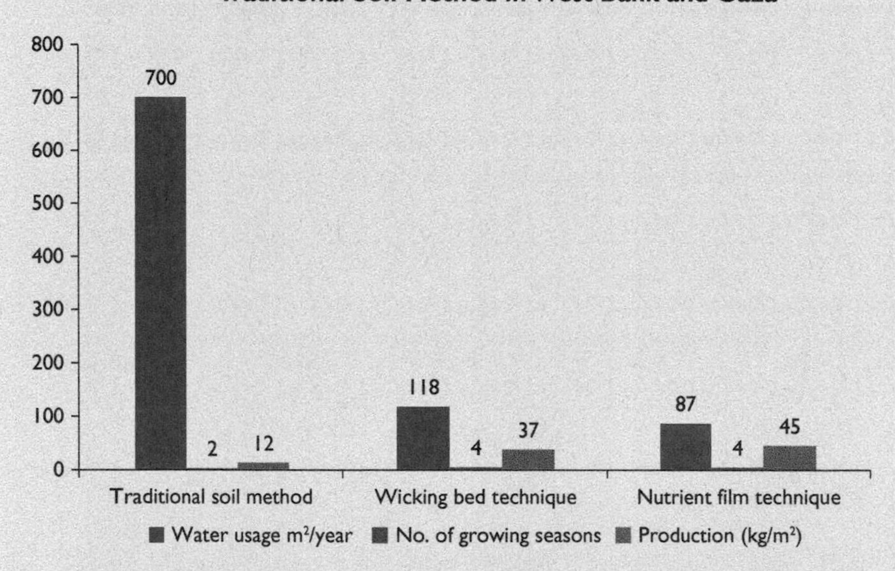

Source: Applied Research Institute–Jerusalem, personal communication, 2016, 2021.
Note: m^2 = square meters.

a. Applied Research Institute–Jerusalem, personal communication, 2016.

in West Bank and Gaza. It shows that NFT uses less water and produces greater crop yields than the wicking bed technique, but both grow crops much more efficiently than soil methods. This is especially the case considering that hydroponics can support three to five growing seasons per year and shorten harvest cycles, depending on crop types. More sophisticated hydroponic variations, such as aquaponics and aeroponics, use even less water than simpler hydroponic systems (Pantanella et al. 2012). The benefits from hydroponic systems compared with rainfed agriculture are not as clear-cut because in rainfed agriculture, rainwater cannot be "wasted" or "saved"—it is merely part of the hydrological cycle.

Growing Medium

A good plant-growing medium should be friable, well drained, well aerated, and moderately fertile. Rockwool is the most widely used hydroponic growing medium, although there are several other popular options, such as sand, gravel, perlite, sawdust, peat moss, vermiculite, and coconut coir. The growing medium's porosity, water holding capacity, water availability, buffering capacity, and cation exchange capacity govern its interaction with the nutrient solution. For example, these factors determine how available nutrients are to the plants, how quickly the nutrient solution passes through the growing medium, and how often irrigation or fertigation—which includes the injection of fertilizers, water amendments, and other water-soluble products into a hydroponic system—is required (Wootton-Beard 2019). Pure topsoil is not recommended as a growing medium for seedlings or transplants because of problems with weeds, disease, drainage, aeration, and inconsistent physical conditions. Specific vegetable seed varieties are needed for certain climates and to resist certain diseases. Other varieties are better adapted to industrial production processes. For example, Roma plum tomatoes grow well in hot climates and are less likely to be damaged during storing, packaging, and transportation.

OUTPUTS

Nutritious Produce

Hydroponic produce has similar amounts of nutrients as conventionally grown produce (Treftz and Omaye 2016). In conventional farming, plants obtain nutrients from the soil, whereas in hydroponics, plants obtain nutrients from a solution. Since plants generate their own vitamins by absorbing nutrients and converting sunlight into energy, there are no differences in vitamin levels among plants grown in soil or in nutrient solutions. However, the mineral content can vary in hydroponic crops depending on the type of nutrient solution used. The nutrient and phytochemical levels differ slightly for all plants, regardless of the growing method. Each crop's nutritional profile depends on the crop's variety, the season in which it is harvested, the length of time between harvest and consumption, and how the crop is handled and stored.

That said, in hydroponic systems, a plant's nutrient levels can be enhanced simply by adding nutrients to the solution. These nutrients could include calcium and magnesium, or microelements such as zinc or iron. Conversely, soil-based systems are relatively inefficient because of nutrient leaching and evaporation, thereby requiring more inputs than hydroponic systems, which can more precisely control the crops' growing environments. As a result, hydroponic systems lead to greater yields, reduced water and nutrient losses, and a greater uniformity of produce.

Different Types of Produce

Many high-input and short-duration crops are grown in hydroponic systems (Wootton-Beard 2019). In principle, it is possible to grow any crop using hydroponics, provided the plant can access enough water and nutrients to support its growth. Hydroponic systems grow many high-input crops such as peppers, tomatoes, strawberries, and cucumbers (Wootton-Beard 2019). The use of hydroponics to grow flowers and other nonvegetable, high-value crops has increased in recent years (Jones 2016). Short-duration crops, which are crops that reach maturity in a short period of time but are sensitive to changes in environmental conditions, are well-suited for hydroponics. These crops include leafy greens, annual herbs, and salad leaves (Wootton-Beard 2019). Short-duration crops require protection from pests and disease and carefully controlled irrigation to maintain leaf quality. Hydroponic systems do this and, in the case of leafy crops, keep them free from soil contaminants (Wootton-Beard 2019). Table 5.1 presents a list of examples of crops grown in hydroponic systems. All these crops tend to be profitable enough to justify a hydroponic operation (Jensen 1999).

TABLE 5.1 Examples of Human Food or Animal Feed from Hydroponic Crops

Vegetables	Fruits	Herbs	Grains/ animal fodder	Flowers
Leafy greens, radishes, celery, cucumbers, potatoes, yams, peppers, wheatgrass, onions, leeks, carrots, parsnips, squash, zucchini, corn, bok choy, kale, Swiss chard, arugula, watercress, chives, broccoli, beans, squash, peas, cauliflower, cabbage, carrots, onions, radishes, beets, microgreens, melon	Tomatoes, watermelon, cantaloupe, strawberries, blackberries, raspberries, blueberries, grapes, dwarf citrus trees (lemons, limes, oranges), dwarf pomegranate tree, bananas	Chives, oregano, mint, basil, sage, rosemary	Barley, oats, wheat, sorghum, alfalfa, cowpea, maize, rice	Roses, peace lilies, hoya, snapdragons, dahlias, carnations, orchids, petunias

Source: Original table for this publication.

Animal Fodder

Hydroponic produce can help meet the world's need for green fodder for livestock. Fodder crops produced by hydroponics are also known as hydroponic fodder, sprouted fodder, or sprouted grain. These crops include oats, wheat, maize, barley, sorghum, cowpea, and alfalfa. The expansion of global livestock populations has significantly increased the world's demand for green animal fodder (Government of India 2010; Brithal and Jha 2010). At the same time, the world's land allocation for green fodder cultivation is limited (Dikshit and Brithal 2010). Hydroponic green fodder is produced from forage grains, such as malt barley or oats, has a high germination rate, and generally takes only seven days to grow (Sneath and McIntosh 2003). Since fodder sprouts are young and tender, they are similar to fresh green grass. They are also high in protein and comprise metabolizable energy; thus, they are highly palatable, highly nutritious, and disease-free.

The hydroponic process to produce green animal fodder is specific. It takes place in an intensive hydroponic growing unit where water and nutrients are used to produce a lush and nutrient-rich grass and root combination. Some systems use tunnels or greenhouses to provide a level of climate control. In greenhouses, water fogging and tube lights automatically maintain light, temperature, and water and humidity levels (Chandra and Gupta 2003). Depending on the type of grain or grass being grown, the forage mats are 15 to 20 centimeters high and produce about 7 to 9 kg of fresh forage, which is equivalent to 0.9 to 1.1 kg of dry matter (Mukhopad 1994; Bustos et al. 2000). The fodder yields are also determined by the seed rate. Most commercial hydroponic units recommend a seed rate of 6 to 8 kg/m^2 (Morgan, Hunter, and O'Haire 1992) for higher outputs (Naik and Singh 2013). Each seed then produces a 200- to 250-millimeter-long vegetative green shoot with interwoven roots in five to eight days. If seed density is high, there is a greater chance of microbial infection in the root mat, which can affect sprout growth.

Green fodder grown from hydroponic systems has various benefits. In the harsh climate in Lodwar town in Kenya, women grew hydroponic barley for feed for smaller livestock, mainly goats. They grew the barley in metal trays with nutrient solution. It took seven days to convert 2 kg of seeds into 12.5 kg of fodder for 10 goats. The goats that consumed the hydroponically grown barley had increased milk yields during the dry season (WFP Kenya 2020). In Chad, more than 230 hydroponic units enabled communities to produce 340 tons of hydroponically grown animal feed in 2020 (Gabe 2020). This was particularly helpful when local feed markets were closed or unsafe. White sorghum is another option for hydroponic animal feed. Various studies estimate that feeding hydroponic fodder to lactating cows increases milk yields by 8 percent (Reddy, Reddy, and Reddy 1988) to 14 percent (Naik and Singh 2013). Farmers in India's Satara district in Maharashtra said that milk production increased by 0.5 to 2.5 liters per animal per day, increasing net profits by 25–50 rupees per animal per day. In addition, cows fed hydroponic fodder had sweeter and whiter milk with increased fat content, improved health and conception rates, and reduced cattle feed requirements by 25 percent. The farmers also said growing the fodder had reduced the cost of labor and required less land than fodder grown on soils (Naik and Singh 2013).

Examples of Animal Fodder Production in West Bank and Gaza

West Bank and Gaza provides an example of successful hydroponic green fodder production. For US$800, farmers in West Bank and Gaza can purchase a hydroponic system optimized for green fodder production. This self-contained system measures 75 × 100 × 180 centimeters and is equipped with seven layers of two growing trays per layer (see photo 5.3). The system is designed to feed goats and sheep on small farms. Two of the 14 trays provide approximately 18 kg of green fodder—enough to feed 7 to 10 sheep or goats each day. The system can maintain this feeding rate for the entire year. The initial investment to establish the system is high for local conditions, but farmers expect to break even within the first year. Moreover, farmers claim that milk production increased by 40 to 50 percent since they implemented their hydroponic fodder systems.[3] Table 5.2 shows the financial breakdown of the system's investment costs.

PHOTO 5.3 Two Views of a Hydroponic Green Fodder System in West Bank and Gaza

TABLE 5.2 Financial Breakdown of the Hydroponic Green Fodder Production System in West Bank and Gaza

Item	Value
Initial investment (US$)	800
System life expectancy (years)	7
Water consumption/year (cubic meters)	18
Water consumption/year (US$)	28
Barley seeds/year (US$)	551
Electricity consumption/year (US$)	85
Operation and maintenance/year (US$)	32
Total annual cost (C)[a] (US$)	811
Total production quantity/year (kilograms)	6,480
Total market value of fodder produced/year (R) (US$)	2,145
Net income (R – C) (US$)	1,334
Gross margin (%)	68

Source: Applied Research Institute–Jerusalem, personal communication, 2021.

a. Initial investment is amortized over the seven-year life expectancy of the system and used in determining total annual costs as well as being part of the cost of goods sold in determining the gross margin.

Examples of Animal Fodder Production in Chad during COVID-19

Refugees fleeing Sudan's Darfur region initiated a successful hydroponic feed production system in 2020 in host communities in the Sahel in Chad. The objective was to improve pastoralists' livelihoods, especially during the dry months of the year when fresh grazing is not available. Led by the WFP, the project benefited 2,770 pastoralists, mostly women, living in the rural host communities of Amnabak, Iridimi, and Touloum in eastern Chad.

The project had two phases that generated many benefits. During the first phase of the project, the pastoralists built hydroponic units that included five vertical shelves per unit and 50 production trays per shelf. The shelves were built by local welders, and the trays were made from discarded jerrycans and other plastic containers. Each unit cost US$400 to build. During the second phase of the project, the beneficiaries used renewable materials, such as bamboo sticks and tree branches, to build the hydroponic units, leading to more cost efficiencies. Such units were free or cost less than US$100 each. The decreased costs led to more pastoralists building units. During the two phases, the beneficiaries installed 554 units. In 2020, these units produced 340 tons of hydroponically grown animal feed, which helped the pastoralists feed their cattle during the dry season. It also helped the pastoralists grow animal fodder when markets were closed or avoided because of the COVID-19 (coronavirus) pandemic. In the end, the project increased milk production; improved meat quality, thereby increasing its market value; and improved the health of the livestock, which decreased the perinatal mortality of the herd.[4]

ADVANTAGES OVER SOIL AGRICULTURE

Higher Yields Than Conventional Agriculture

The hydroponic system is more productive and efficient than conventional farming. Hydroponic crops have greater yields and, often, higher nutritional values than traditional agricultural crops (Christie 2014; Buchanan and Omaye 2013; Gichuhi et al. 2009; Selma et al. 2012; Sgherri et al. 2010). Hydroponic systems also require fewer pesticides (Resh and Howard 2012). The reason for the greater yields is that hydroponics allows for continuous year-round production and shorter harvest cycles than soil-based farming methods. The exact number of annual growing seasons depends on the type of hydroponic system being used and the climatic conditions within that system. For example, outdoor system growing seasons are still dictated by the outdoor temperature and daylight hours. Plants grown hydroponically are generally less stressed than soil-grown plants since hydroponic plants are in their optimum growing conditions all the time, which creates less waste than conventional farming (Treftz and Omaye 2016). Moreover, hydroponic crops have similar, and sometimes higher, levels of vitamins and minerals compared with soil-based cultivation systems (Gruda 2009). The pilot in Kenya's Kakuma refugee camp (box 5.1) showed that hydroponic systems require up to 75 percent less space than traditional field farming methods (WFP Kenya 2020).

Reduced Water Usage

Hydroponic systems use less water than open-field agriculture (Ly 2011; Despommier 2010). According to Despommier (2010), hydroponics requires approximately 80 to 99 percent less water than traditional agriculture, with the more advanced hydroponic systems using less water than simplified systems. Hydroponic and aeroponic techniques deliver the optimal amount of water needed for healthy plant growth. The same principle is true for nutrients, which producers can mix precisely and deliver to plants, thereby optimizing the growing conditions for each plant's species, growth stage, and nutrient requirements (Wootton-Beard 2019).

Hydroponic systems have been successful in water-stressed environments in Africa. Between October 2018 and March 2019, a hydroponic pilot project was carried out in Djibouti (Ministry of Agriculture of Djibouti 2019), one of Africa's more arid countries. The project benefited 27 Djiboutians. In Kenya, Hydroponics Africa Limited provides training and installations of hydroponic, aquaponic, and greenhouse structures. The company has installed more than 3,500 hydroponic units and trained more than 5,000 small-scale farmers in Rwanda, Somalia, Tanzania, and Uganda. Sixty-five percent of these units are located in dry regions. The WFP hydroponic pilot project in the Kakuma refugee camp (box 5.1) showed that hydroponics decreased water usage for kale by 82 percent, spinach by 92 percent, and cowpeas by 84 percent compared with conventional farming (WFP Kenya 2020). It also showed that hydroponics uses 1.4 to 6 times less space, has greater yields, and has shorter growing cycles than conventional farming (figure 5.5).

Versatility

Hydroponic farming is possible across diverse climates and agro-ecological zones, including arid areas and urban zones (Heredia 2014). This is because hydroponics can be applied indoors. Growing in greenhouses or other controlled environments separates the production area from the location's natural ecosystem; therefore, the ecosystem has no impact on the growth of hydroponic plants. As such, hydroponic farming can be done anywhere. This separation of growing from the natural environment also eliminates any environmental harm that agriculture would cause to the natural ecosystem, such as deforestation, monoculture, or any other form of environmental degradation. Thus, hydroponic food production has a minimal impact on natural resources and the environment and can be implemented in cities or on degraded lands.

Producing crops in urban areas minimizes the distance between the food producer and the urban consumer (Bellows, Brown, and Smit 2004). A closer proximity of producers to markets reduces labor, transportation, packaging, and refrigeration requirements, leading to potentially substantial decreases in the use of resources and energy. In the United States, these additional costs account for up to 79 percent of a crop's retail price (Wohlgenant 2001).

FIGURE 5.5 Hydroponic Space, Water Needs, and Yields for Producing Kale, Spinach, and Cowpeas

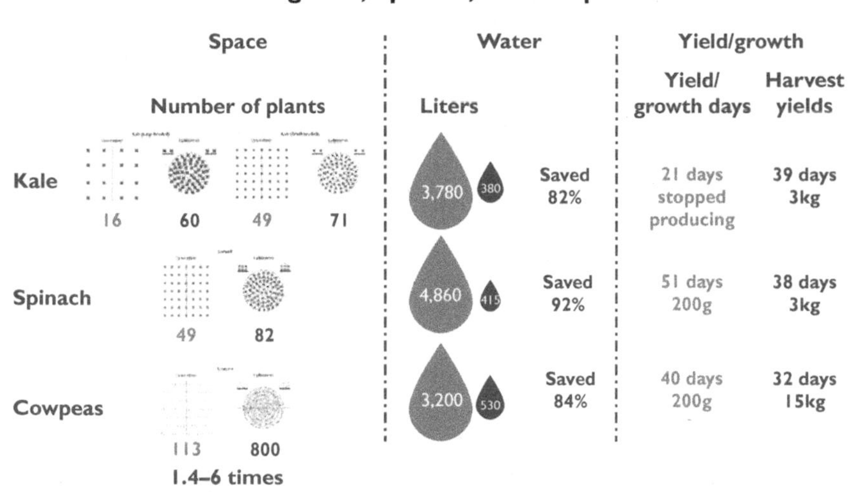

Source: WFP Kenya 2020.
Note: g = grams; kg = kilograms.

Shortening and simplifying food supply chains can drastically diminish their environmental impacts, while providing cities and rural areas fresh, highly nutritious produce. A pilot hydroponic project in The Gambia demonstrated some of the benefits of reducing producer-to-consumer distances (Abdoulaye 2009). The project focused 80 percent of its efforts in urban and peri-urban areas, where poverty rates can exceed 40 percent.

Hydroponic systems can be set up in small spaces, such as homes, and large spaces, such as commercial farms. Hydroponic systems are controlled and can be built in any environment. Hydroponic farms have been established in unused or recycled spaces such as parking lots, building rooftops, shipping containers, abandoned warehouses, and even underground tunnels such as sewers or subways. The type of space depends on the type of crop being grown. Hydroponic farmers tend to use smaller spaces to grow herbs and leafy greens because these plants grow quickly, can be continuously harvested, and do not require much space to expand. Therefore, growing herbs and leafy greens in warehouse facilities that are vertically oriented requires little space but can generate lots of produce. Hydroponic farmers often use larger spaces for voluminous plants that require more advanced hydroponic systems with trellises or deep root support. In 2018, the largest hydroponic tomato greenhouse in West Africa, Wells Hosa Greenhouse Farms, opened in Nigeria. Located on 27 hectares of land, it consists of 28 hydroponic greenhouses that are each 5,440 m^2. The company aims to produce 4,200 metric tons of tomatoes to meet local and export demand and generate US$6 million in annual income.

Simple hydroponic systems do not require much labor. Reuters tells the story of Venensia Mukarati, a Zimbabwean woman who was able to set up and operate her own hydroponic system. Ms. Mukarati did this because she wanted to grow produce for her family but did not have land. Through a simple online search, she learned how to grow vegetables on her deck using a small hydroponic system that she imported from Cape Town for US$900. Ms. Mukarati quickly realized that hydroponics could be a profitable venture. She learned that she could harvest vegetables within six weeks with hydroponics instead of 10 weeks if she were using conventional soil-based agriculture. She started by growing 140 plants but now produces 2,600 plants per cycle, including herbs, lettuce, spinach, and cucumbers in two makeshift greenhouses. After two years, she scaled up her production fourfold by building larger greenhouses on 2,600 m^2 of land. By January 2020, Ms. Mukarati was also training others in hydroponics and selling a hydroponic "starter pack" for US$200 that she designed herself. She earns US$1,100 per month, or about 14 times what some government workers earn. She established this system on her own and requires little help to maintain its operations (Dzirutwe 2020).

Pest Management

Hydroponic plants are usually more pest resistant than soil-based plants and do not need chemical herbicides or pesticides. Pest infestations can destroy crops, and pesticide use is a health and environmental concern for many consumers. In hydroponic farming, pesticide use is usually unnecessary. A study by Treftz and Omaye (2016) compares strawberry growth performance in hydroponic and soil systems. It shows that 80 percent of hydroponic strawberries survived compared with only 50 percent of soil-grown strawberries. The study cites pest infestations as the cause of the lower plant survival rates in soil-based farming (Treftz and Omaye 2016). This suggests that using hydroponic systems on a large scale could reduce pesticide usage, reducing the farmer's input costs.

Energy Efficiency

Hydroponics is feasible in areas with unreliable seasonal sunlight. Several studies have shown that both light intensity and light quality are important for plant growth and development (Neff, Fankhauser, and Chory 2000; Fukuda et al. 2008). Plants use light for photosynthesis and to signal the start and end of the growing season. Artificial light can replace sunlight in indoor plant growing systems. However, reliance on artificial light increases electricity use and energy costs.

Hydroponics can reduce carbon emissions if the system uses renewable energy sources or natural heating. Addressing electricity needs is one of the key trials facing the hydroponic industry, particularly in northern latitudes. High-tech hydroponic systems tend to use a lot of energy since they usually incorporate lighting, pumping, heating, and air moderation systems. Producers can mitigate this by locating hydroponic facilities in areas with access to inexpensive renewable sources of energy, such as wind, solar, or geothermal power (Barbosa et al. 2015). Increased greenhouse gas emissions from high energy use are partially offset by reduced transportation needs. Using greenhouses to store natural heat energy can also reduce energy needs. Greenhouses located in more moderate climates, such as climates closer to the greenhouse set point temperature, will experience a lower energy demand. In certain climates, heating and cooling systems may not be required but instead replaced by a passive ventilation system, thus reducing the overall energy demand considerably (Barbosa 2015).

Benefits for African and FCV Countries

Hydroponic technologies are climate-resilient and mitigate climate-related risks. Climate change is exacerbating fragility in FCV countries (World Bank Group 2019). In East and Southern Africa in 2020 and 2021, a hotter climate was linked to swarms of locusts infesting subsistence crops, especially affecting Kenya's nearly 9 million smallholder farmers (World Bank Group 2019). Hydroponics, which can be established anywhere in climate-controlled conditions, is not exposed to temperature variations and can hedge against the risk that climate change and secondary effects, such as pests, pose to traditional field crops.

Hydroponics can support youth populations. In Djibouti, for example, youth are being educated on the basics of hydroponic technology, making agriculture more appealing for them. In Kenya, hydroponic technologies are being integrated into higher education with the Kenya Education Management Institute, an agency of Kenya's Ministry of Education, which has established a demonstration center for hydroponic farming. In South Africa, the Makotse Women's Club, in partnership with the Ministry of Agriculture, is training unemployed youth, AIDS orphans, and other youth community members to start their own hydroponic businesses. The training also covers HIV prevention, nutrition, and environmental conservation.

Hydroponics can support women and other vulnerable populations, such as internally displaced persons. An illustrative example of this is from the northern Darfur refugee camp Zamzam. A German nonprofit, Welthungerhilfe, partnered with a Sudanese nongovernmental organization, Al Rayan for Social Development, to launch a hydroponic project in the camp (Welthungerhilfre 2020). By 2020, they were piloting this project with 150 women who were using locally available materials to build basic hydroponic systems, called "set it and forget it" methods. These intelligent drip irrigation systems use smart tubing to regulate irrigation to the plants. Welthungerhilfe links the internally displaced women with local banks and microfinance institutions to set up village savings and loan associations. As such, project beneficiaries can obtain loans to continue the hydroponic systems on their own. The Sudanese Ministry of Agriculture provides technical support to the beneficiaries, and local agricultural suppliers provide certain inputs. Figure 5.6 shows how hydroponics can build peace in FCV countries.

LIMITATIONS

Poorly implemented hydroponic systems are vulnerable to failure. In systems where roots are highly exposed, plants can dry out rapidly. And in hydroponic solutions, nutrient and pH imbalances can build up far more quickly than in soil. Likewise, waterborne diseases and microorganisms, harmful bacteria, and damaging rots and molds can spread quickly and widely, contaminating solutions fairly easily. These risks are greater in recirculating systems where pathogens can build up over time. Therefore, if something goes wrong in a hydroponic system, entire crops can be wiped out very quickly. Farmers can reduce these risks by following proper sanitation measures such as effective integrated pest management and regular testing and treatment of irrigation water (Wootton-Beard 2019). Covered crops can minimize these risks. And effective identification and treatment schedules are necessary for fungal diseases, which are more likely to occur in the warm, humid conditions of hydroponic systems.

FIGURE 5.6 How Hydroponics Supports the World Bank Group's Four FCV Pillars

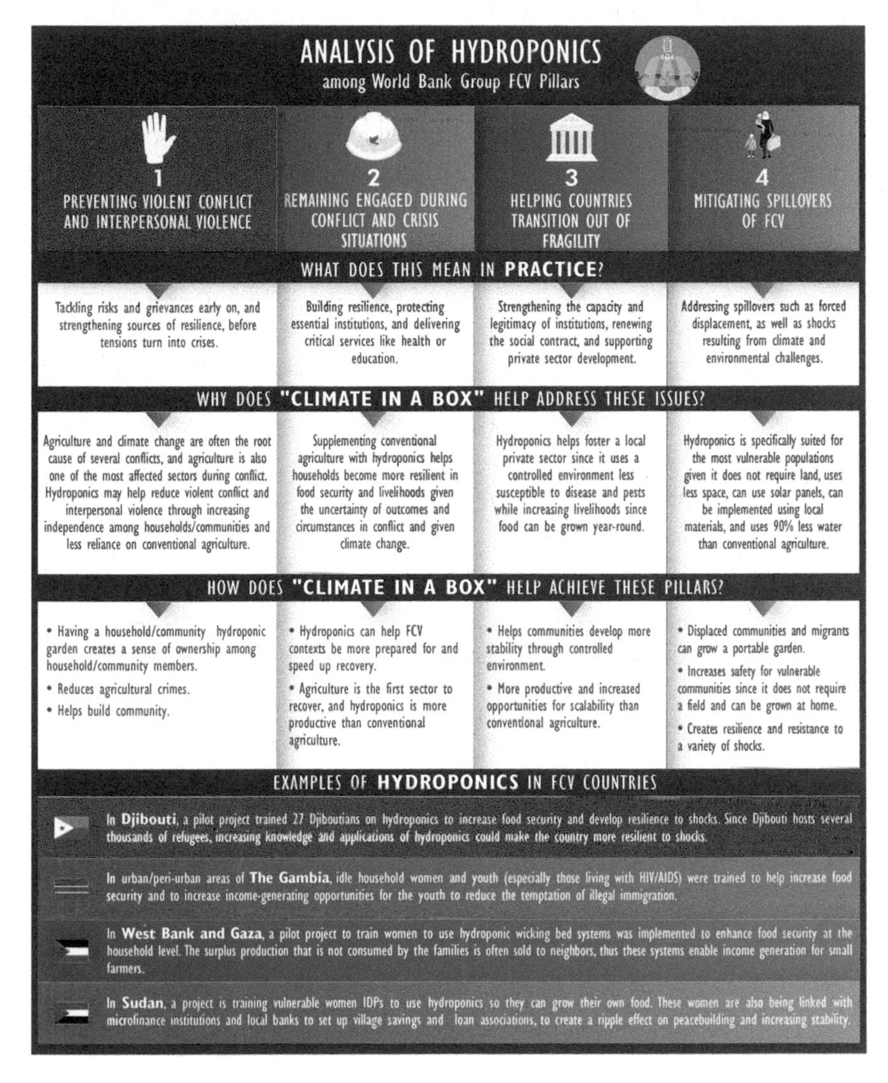

Source: World Bank Group 2019.
Note: FCV = fragility, conflict, and violence; IDPs = internally displaced persons.

Economic Viability: Costs, Labor, and Profitability

Costs

As hydroponic systems are scaled up, production costs decrease and profits come sooner. In the pilot in the Kakuma refugee camp (box 5.1), the first year's investment cost for one household unit to install six circular gardens with the required inputs, such as seedlings and nutrient water solutions, was US$1,006: US$926 for the hydroponic structure and US$80 annually for inputs. Households used the hydroponic produce for additional income and household consumption. Three circular gardens would allow a single household to consume vegetables at least twice a week compared with only once a week previously. Households could sell excess produce to generate additional income. In one year, the household unit earned US$173 when it kept half the harvest from the six units for self-consumption and US$347 when it sold the entire harvest. By this measure, it would take a household 3.2 years to recoup the investment costs if it sold the entire harvest, and nine years if it sold half and kept the other half for personal consumption. The breakeven point was even faster for group units of 40 circular gardens that supplied 10 beneficiaries. The first year's investment for the group was US$5,740: US$5,140 for the structure and US$600 for annual variable inputs. The group unit earned US$2,469 when it sold its entire annual harvest. At this rate, the breakeven point for a group unit was 2.4 years, or 3.6 years if the beneficiaries kept 25 percent of the harvest to consume themselves.

Simple low-tech hydroponic equipment costs are relatively low for areas where imports are expensive and local resources are available. Commercial hydroponics requires an expensive greenhouse or warehouse and specialized equipment, depending on the complexity of the system being used. Some commercial hydroponic operations also require controllers, computer systems, large-scale lighting fixtures, ventilation and heat recovery systems, irrigation and rainwater harvesting, and specialized labor (Pantanella et al. 2012). For a basic wick system in West Bank and Gaza, a local wicking bed unit kit costs US$820. The kit contains four beds and, depending on crop type, can plant up to 100 seedlings per season for three to five seasons per year. The average annual yield is 156 kg, with an average total market value of US$220.[5] The cost of an NFT hydroponic system in West Bank and Gaza is US$1,000, and the system can yield 175 kg of produce, with a total market value of US$245.[6] The installation cost of a hydroponic pipe system package, a third system available in West Bank and Gaza, is US$1,135. The package includes a shading system, which consists of a shadow mesh held up by a metal frame, and an electrical conductivity meter, pH meter, timer, and submersible pump. The package consists of eight pipes, each with 16 holes—or planting eyes—per pipe. Thus, the system allows for 128 seedlings per planting cycle (photo 5.4). These simplified hydroponic systems can become profitable, but the start-up costs are high for poor farmers. Simplified hydroponic systems cost significantly less if labor and materials are locally available. The West Bank and Gaza example shows that costs are relatively low for simple systems in areas where imports are expensive and local resources are available. For a comparable example, tables 5.3 and 5.4 show the variable and total costs, respectively, associated with hydroponic cucumber production in Turkey.

Hydroponic greenhouses can cost 2 to 20 times more than soil agricultural systems (Mathias 2014). These costs can increase even more with sophisticated

PHOTO 5.4 Hydroponic Farming in West Bank and Gaza

a. Tomatoes growing in hydroponic pipes

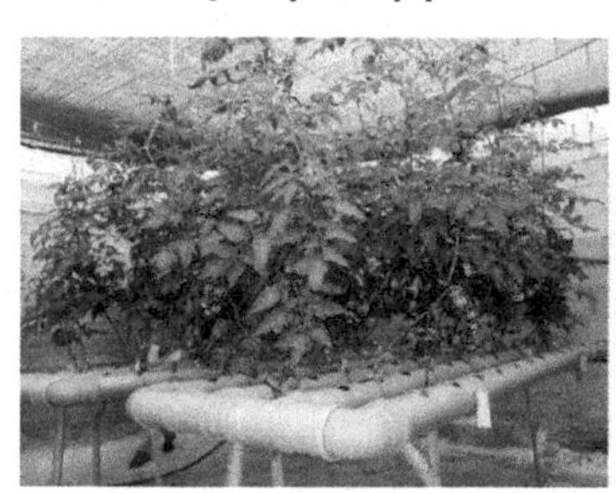

b. Training schoolgirls in West Bank and Gaza in hydroponic pipe farming

TABLE 5.3 Variable Costs of a 1,035 Square Meter Hydroponic Cucumber Production Operation in Turkey

Operation	Total cost ($)	Share of cost (%)
Growing medium preparation (labor) (42 hours)	65	3.9
Fertilizing (labor) (18 hours)	32	1.9
Calcium nitrate (82 kg)	69	4.1
Ammonium nitrate (17 kg)	46	2.8
Potassium nitrate (163 kg)	127	7.6
Magnesium sulfate (84 kg)	81	4.8
Iron (4 kg)	74	4.4
Phosphoric acid (46 l)	76	4.5
Nitric acid (66 l)	58	3.4
Micro element (48 l)	47	2.8
Planting (labor) (32 hours)	28	1.7
Seedling (3,150 seeds)	354	21.1
Irrigation (labor) (35 hours)	36	2.1
Electricity (98 kWh)	45	2.7
Pruning and training (labor) (40 hours)	51	3.0
Fungicide and insecticide application (labor) (4 hours)	38	2.3
Fungicide and insecticide	174	10.4
Yellow trap (5 kg)	12	0.7
Glue (for traps) (4 tubes)	10	0.6
Harvest (labor) (270 hours)	98	5.8
Wrapping (labor) (76 hours)	41	2.4
Plastic wrapping (13 kg)	32	1.9
Transport	85	5.1
Total variable costs	**1,679**	**100.0**
Per square meter	**1.62**	

Source: Engindeniz 2004.

Note: kg = kilograms; kWh = kilowatt-hours; l = liters.

TABLE 5.4 Total Costs of a 1,035 Square Meter Hydroponic Cucumber Production Operation in Turkey

Item		Total cost ($)	Share of cost (%)
Variable costs (1)		1,679	42.4
Fixed costs (2)	Interest on total initial investment costs	1,050	26.5
	Annual initial investment costs	911	23.0
	Interest on total variable costs	118	3.0
	Administrative costs	50	1.2
	Total	2,129	53.7
Land rent (3)		154	3.9
Total costs (1+2+3)		**3,962**	**100.0**
Per square meter		**3.83**	

Source: Engindeniz 2004.

hydroponic food production methods, such as aeroponics. A commercial greenhouse that measures 279 m^2 with complete heating, cooling, and ventilation systems costs between US$10,000 and US$30,000. Low-cost greenhouses, such as hoop houses and attached solar greenhouses, can be constructed for as little as US$500 to US$1,500 (Greer and Diver 2000). A modern greenhouse with a hydroponic plant growing system costs US$90 to US$100/m^2 to build, excluding the cost of land. Glass panels for greenhouses can cost as much as US$140/$m^2$ (Jensen and Malter 1995). Building multiple greenhouses would increase the total expenditure but reduce the cost per square meter (Engindeniz 2009). Table 5.5 shows the initial investment costs for greenhouse construction in Turkey.

Labor

Labor and energy are the main operating costs in colder climates. The labor costs can be mitigated in developing countries, where labor is much cheaper, and by using automated systems, which reduces the system's reliance on manual labor. Moreover, hydroponics requires less labor after the system's initial installation (Daly and Fink 2013).

People without a formal education can rapidly acquire the skills to operate hydroponics. Soil-grown produce has increased labor costs because of weeding, watering, and pesticide spraying requirements (Resh and Howard 2012). Field visits showed that labor costs are higher in more sophisticated hydroponic systems, which require more technical expertise to monitor and troubleshoot when problems arise. In industrial-sized hydroponics, growers must have technical skills in crop species, plant health, nutrient deficiency and toxicity, nutrient solutions, electricity, and water circulation systems. Therefore, it is

TABLE 5.5 Initial Investment Costs for Constructing a 23 x 45 Meter Greenhouse in Turkey

Item	Initial cost ($)	Share of cost (%)	Useful life (years)	Annual cost ($)	Share of cost (%)
Galvanized frame and kit	4,552	30.3	20	455	25.0
Base locking rail	3,912	26.1	20	196	10.8
Polyethylene covering (250 kg)	769	5.1	2	385	21.1
Ground covering	280	1.9	2	140	7.7
Polyethylene bed (323 kg)	686	4.6	5	137	7.5
Volcanic tuff	523	3.5	5	105	5.8
Water pipe (PVC) (125 m)	183	1.2	15	12	0.7
Water pipe (dropping) (1,500 m)	249	1.7	10	25	1.4
Plastic filter	50	0.3	10	5	0.3
Plug	18	0.1	10	2	0.1
Nipple	19	0.1	10	2	0.1
Elbow (PVC)	30	0.2	15	2	0.1
Valve	45	0.3	15	3	0.2
Adapter	18	0.1	10	3	0.2
Record	10	0.1	10	1	0.1
Hose-pipe	11	0.1	10	1	0.1
Lapel	18	0.1	10	2	0.1
pH meter	212	1.4	10	21	1.1
Electric conductivity meter	199	1.3	10	20	1.1
Thermic button	28	0.2	10	3	0.2
Timer	132	0.9	15	9	0.5
Buoy	10	0.1	10	1	0.1
Water and fertilizer tank	62	0.4	15	4	0.2
Water pump (12 m³/hour)	330	2.2	15	22	1.2
Site preparation and ground gravel	829	5.5	—	83[a]	4.5
Assembly and installation	1,830	12.2	—	183[a]	10.0
Total	**15,005**	**100.0**	—	**1,822**	**100.0**
Per square meter	**14.50**	—	—	**1.76**	—

Sources: Engindeniz 2004, 2009.

Note: kg = kilograms; m = meters; PVC = polyvinyl chloride; — = not applicable.

a. Calculated over 10 years.

necessary to hire qualified employees to manage industrial hydroponic systems (Corrêa et al. 2012). Examples from Jordan, the United Arab Emirates, and West Bank and Gaza show that people without a formal education can rapidly acquire hydroponic skills (Verner et al. 2017). Agricultural extension services can provide evidence-based educational tools in clear language to train farmers in these skills and business owners in the economic and environmental benefits of hydroponics (Treftz and Omaye 2016).

The type and amount of employment generated with hydroponics depend on several factors: (1) the type of hydroponic system, with simple systems requiring less labor than sophisticated systems; (2) the size of the system since small-scale systems for domestic use require less labor than large commercial operations; (3) the level of worker expertise, because there will be needs for low-skilled workers who perform multiple operations and high-skilled workers with more technical expertise; (4) the type of crops, because, for example, leafy greens require more labor than most fruit producing plants; (5) the harvesting and marketing process, because, for example, packaging a whole head of lettuce in a plastic bag requires less labor than picking leaves from several heads to create mixed leaf bags; and (6) the market, because a more developed market will require less labor to sell surplus produce. Based on field observations, a single wicking bed unit used for personal consumption can provide one part-time job. Women who maintain such single-bed systems need to work only two or three hours a day for two or three days per week.[7] It is difficult to obtain employment information for larger-scale commercial operations because their data are private. However, evidence shows that using a deep water culture or NFT system to grow leafy greens on 1 acre of land provides approximately 18 to 22 full-time jobs.[8]

Profitability

Despite their high costs, hydroponic systems can be profitable. In The Gambia, hydroponic production costs are US$2.30 for 1 m^2 of lettuce and US$3 for 1 m^2 of sweet pepper, whereas profits reach US$6 for lettuce and US$15 for sweet pepper. This corresponds to returns of US$2.60 for lettuce and US$5 for sweet pepper for every dollar invested (Abdoulaye 2009). Returning to the example of hydroponic cucumbers in Turkey, the total gross revenue obtained from selling 25,384 kg of cucumbers grown in a 1,035 m^2 hydroponic system was US$5,839 (US$5.64/m^2). Table 5.6 shows the profitability for this example. The simplified hydroponic systems in West Bank and Gaza also showed profits. Table 5.7 compares two of the hydroponic systems available for purchase in West Bank and Gaza: a 10 m^2 wick system with 10 wicking beds and the eight-pipe hydroponic pipe system. Farmers earned profits from both systems, although the wicking beds show a slight advantage in the overall gross margin.

A study from Kenya showed that hydroponics is profitable under certain circumstances (Croft 2016). Profitable circumstances include when the system grows high-value crops; is set up in areas where input costs are low, water constraints are high, and arable land is scarce; and is located closer to consumers

TABLE 5.6 Net Financial Returns Obtained from a 1,035 Square Meter Hydroponic Cucumber Production Operation in Turkey

Item	Total (US$)	Proportion of revenue (%)
Total gross revenue (R)	5,839	100.0
Variable costs	1,679	28.8
Fixed costs	2,129	36.5
Land rent	154	2.6
Total costs (C)	3,962	67.9
Total net return (R) – (C)	**1,877**	**32.1**
Per square meter	**1.81**	

Source: Engindeniz 2004.

TABLE 5.7 Profitability of Two Hydroponic Systems in West Bank and Gaza

Item	System	
	Hydroponic pipe	Wicking bed
Initial investment (US$)	1,100.00	1,687.00
Installation cost (US$)	35.00	85.00
Number of planting cycles/year	3 to 5	3 to 5
Seedlings/planting cycle[a]	128	160 to 250
System life expectancy (years)	10	12
Water consumption/year (m^3)	3.1	4.5
Fertilizers and seedlings/year (US$)	65.00	180.00
Other costs (water, electricity)/year (US$)	55.00	85.00
Operation and maintenance/year (US$)	50.00	28.00
Total annual cost (C)[b] (US$)	283.50	440.70
Total production quantity (kg/year)	245	395
Total market value of production/year (US$) (R)	374.60	604.00
Net income (R – C) (US$)	91.10	163.30
Gross margin (%)	24	27

Source: Applied Research Institute–Jerusalem, personal communication, 2021.
Note: kg = kilograms; m^3 = cubic meters.
a. The number of seedlings per planting cycle for hydroponic pipes is limited by the number of planting holes (16 per pipe; 128 per eight pipes); the number of seedlings per planting cycle for wicking beds depends on crop type.
b. Initial investment and installation costs are amortized over the life expectancy of each system and are used in determining total annual costs as well as being factored into the cost of goods sold in determining each gross margin.

than other vegetable markets. The examples described in the chapter, in Turkey and West Bank and Gaza, describe profitable operations. Other cost-benefit analyses of hydroponic systems have focused on high-value crops, such as flowers (Grafiadellis et al. 2000) or melons (Shaw et al. 2007), and found positive results compared with traditional production systems. A study conducted in the western Kenyan town of Eldoret, however, shows that hydroponic systems are not profitable under all circumstances (Croft 2016). The study evaluated three hydroponic systems—Kratky, NFT, and ebb and flow—growing African leafy vegetables (ALV) for profitability and nutrient concentration, with the ebb and flow system only producing seedlings for the other two hydroponic systems. The study calculated the net present value (NPV) and benefit-cost ratio for each system with a five-year time horizon. However, neither the soil-based system nor any of the hydroponic systems was profitable compared with purchasing ALV at the market. Soil-based production would become profitable only if the opportunity cost was close to zero. Among the three hydroponic systems analyzed, the NFT system was the closest to profitable, while the Kratky system consistently had the worst benefit-cost ratio and NPV values. For hydroponic systems to be profitable, vegetable prices would have to increase by 1,027 percent or input costs would have to fall below zero. However, ALV are not considered high-value crops, which may explain some of the more negative results. The hydroponic yields were also much lower than yields from soil-based systems (table 5.8), even when considering the extended growing season that hydroponic systems can offer. The hydroponic production systems become more viable and, therefore, profitable when alternative vegetable sources become costlier—as is more likely the case for high-value horticulture—or when neither arable land nor irrigation is available. Under the average conditions found in western Kenya, however, the study found that none of the hydroponic systems was profitable or competitive compared with soil-based production (table 5.9) and purchasing ALV from markets was the most cost-effective alternative.

Processing hydroponic crops can improve their profitability. Value-added products comprise raw produce that has been modified or enhanced to have a higher market value and a longer shelf life, such as dried fruit. For example,

TABLE 5.8 Time, Harvest, and Input Costs for Three Hydroponic Systems and Comparison with Soil-Based Production

Assumed value	Kratky	NFT	Ebb and flow	Soil
Time (days/year/m^2)	1.97	0.75	0.97	1.89
Harvest (g/year/m^2)	530	511	486 seedlings	865
Annual input costs (K Sh/year/m^2)	2,044	607	1,534	0.39

Source: Croft 2016.

Note: 100 Kenya shillings (K Sh) = 0.99 US$; g = grams; m^2 = square meters; NFT = nutrient film technique.

TABLE 5.9 Net Present Value and Benefit-Cost Ratio for Three Hydroponic Systems Compared with Soil-Based Production and Purchasing Vegetables with a Five-Year Horizon and 1 and 10 Percent Discount Rates

Item	Soil-based production alternative				Purchasing alternative		
	Kratky	NFT	Ebb and flow	Soil	Kratky	NFT	Ebb and flow
1% discount rate							
Net present value (K Sh)	−13,011	−2,526	−9,640	−3,932	−13,502	−3,467	−8,262
Benefit-cost ratio	0.24	0.64	0.41	0.08	0.21	0.49	0.50
Time cost (K Sh)	−54,561	822	2,502	5	−61,27	1,542	4,523
Price (K Sh/g)	−7,9	−1,4	5,1	1.0	2,7	0.8	2.8
Harvest (g)	54,904	22,565	3,709	141,183	28,869	15,972	2,006
Inputs (K Sh)	−955	261	−678	−9,712	−1,107	−321	−252
10% discount rate							
Net present value	−10,620	−2,253	−8,292	−3,080	−11,012	−2,997	−7,224
Benefit-cost ratio	0.25	0.77	0.47	0.11	0.22	0.59	0.57
Time cost (K Sh)	−38,158	887	2,775	4	−6,054	1,689	5,300
Price (K Sh/g)	−8.3	−1.6	7,2	1.0	2.8	0.8	3.0
Harvest (g)	57,338	25,549	4,010	141,566	30,110	17,512	2,159
Inputs (K Sh)	−1,136	39	−980	−9,740	−1,291	−550	−557

Source: Croft 2016.

Note: Breakeven values for time cost, price of vegetables or seedlings, harvest, and inputs are also given. 100 Kenya shillings (K Sh) = 0.99 US$; g = grams; NFT = nutrient film technique.

processors can convert basil into pesto, tomatoes into tomato paste, cucumbers and peppers into pickled vegetables, and strawberries into jam, and blueberries can be dried and sold as an antioxidant snack or mixed into cereals. In addition, processors can convert herbs, such as mint, basil, and oregano, into essential oils. This processing can increase the value of high-value crops, such as tomatoes and chili peppers, even further, thereby increasing crop profitability (WFP Kenya 2020).

NOTES

1. infarm.com.
2. Applied Research Institute–Jerusalem, personal communication, 2016.
3. Applied Research Institute–Jerusalem, personal communication, 2021.
4. This information was received through discussions with the WFP's project leaders in Chad. For more information, see Gabe (2020).

5. Applied Research Institute–Jerusalem, personal communication, 2016.
6. Applied Research Institute–Jerusalem, personal communication, 2016.
7. Applied Research Institute–Jerusalem, personal communication, 2017.
8. Dr. Merle Jensen, personal communication, 2017.

REFERENCES

Abdoulaye, S. 2009. "Developing a Soilless Vegetable Production Technology as a Component of Poverty Reduction Strategies in Urban and Periurban Areas of The Gambia." In *9th African Crop Science Conference Proceedings*, edited by S. Abdoulaye, J. S. Tenywa, G. D. Joubert, D. Marais, P. R. Rubaihayo, and M. P. Nampala, 87–91. Cape Town, South Africa, September 28–October 2.

Agritecture. 2019. "Funding for Vertical Farming Startups Is Increasing." Agritecture, New York. https://www.agritecture.com/blog/2019/7/18/funding-for-vertical-farming-startups-is-increasing.

Arnon, D. I., and P. R. Stout. 1939. "The Essentiality of Certain Elements in Minute Quantity for Plants with Special Reference to Copper." *Plant Physiology* 14 (2): 371–75.

Barbosa, G. L., F. D. A. Gadelha, N. Kublik, A. Proctor, L. Reichelm, E. Weissinger, G. M. Wohlleb, and R. U. Halden. 2015. "Comparison of Land, Water, and Energy Requirements of Lettuce Grown Using Hydroponic vs. Conventional Agricultural Methods." *International Journal of Environmental Research and Public Health* 12 (6): 6879–91.

Bellows, A. C., K. Brown, and J. Smit. 2004. "Health Benefits of Urban Agriculture." Community Food Security Coalition's North American Initiative on Urban Agriculture, Portland, OR. http://www.co.fresno.ca.us/uploadedfiles/departments/behavioral_health/mhsa/health%20benefits%20of%20urban%20agriculture%20(1-8).pdf.

Bregnballe, J. 2015. *A Guide to Recirculation Aquaculture: An Introduction to the New Environmentally Friendly and Highly Productive Closed Fish Farming Systems*. Rome: Food and Agriculture Organization of the United Nations. http://www.fao.org/3/i4626e/i4626e.pdf.

Brithal, P.S., and A.K. Jha. 2010. "India's Livestock Feed Demand: Estimates and Projections." *Agricultural Economics Research Review* 23: 15–28.

Buchanan, D. N., and S. T. Omaye. 2013. "Comparative Study of Ascorbic Acid and Tocopherol Concentrations in Hydroponic- and Soil-Grown Lettuces." *Food and Nutrition Sciences* 4 (10): 1047–53.

Bustos, C. D. E., E. L. Gonzalez, B. A. Aguilera, and G. J. A. Esptnoza. 2000. "Forraje Hidropónico, Una Alternativa para la Suplementación Caprina en el Semidesierto Queretano." Reunión Nacional de Investigación Pecuaria 38: 383. Puebla, México.

Campbell, N. A., and J. B. Reece. 2002. *Biology*, sixth ed. San Francisco, CA: Benjamin Cummings.

Chandra, P., and M. J. Gupta. 2003. "Cultivation in Hi-Tech Greenhouses for Enhanced Productivity of Natural Resources to Achieve the Objective of Precision Farming." In *Precision Farming in Horticulture*, edited by J. Singh, S. K. Jain, L. K. Dashora, and B. S. Chundawat, 64–74. New Delhi, India: New India Publishing Agency.

Christie, E. 2014. "Water and Nutrient Reuse within Closed Hydroponic Systems." Master's thesis, Georgia Southern University, Statesboro, GA.

Clark, S. 2018. "Vertical Farming Funding on the Rise in 2017 & Predictions for 2022." Agritecture, New York. https://www.agritecture.com/blog/2017/12/29/vertical-farming-funding-on-the-rise-in-2017-predictions-for-2022.

Corrêa, R. M., S. I. do Carmo Pinto, E. S. Reis, and V. A. M. de Carvalho. 2012. "Hydroponic Production of Fruit Tree Seedlings in Brazil: A Standard Methodology for Plant Biological Research." In *Hydroponics: A Standard Methodology for Plant Biological Researches*, edited by T. Asao, 225–44. IntechOpen. http://www.intechopen.com/books/hydroponics-a-standard-methodology-for-plant-biologicalresearches/production-of-fruit-seedlings-in-hydroponics.

Croft, M. 2016. "The Role of African Leafy Vegetables in Food Security." PhD thesis, Purdue University, West Lafayette, IN. https://docs.lib.purdue.edu/open_access_dissertations/918.

Daly, W., and J. Fink. 2013. "Economic Assessment of Hydroponic Lettuce Production." Goucher College, Baltimore, MD.

Despommier, D. 2010. *The Vertical Farm: Feeding the World in the 21st Century*. New York: St. Martin's Press.

Devictor, X., and Q. Do. 2017. "How Many Years Have Refugees Been in Exile?" *Population and Development Review* 43 (2): 355–69. https://documents1.worldbank.org/curated/en/549261472764700982/pdf/WPS7810.pdf.

Dikshit, A. K., and P. S. Brithal. 2010. "India's Livestock Feed Demand: Estimates and Projections." *Agricultural Economics Research Review* 23: 15–28.

Dzirutwe, M. 2020. "In Zimbabwe, a Backyard Hydroponic Farm Beats Drought to Grow Vegetables." Screenocean, Harare, Zimbabwe. https://www.reuters.com/article/us-climate-change-zimbabwe-hydrofarm/in-zimbabwe-a-backyard-hydroponic-farm-beats-drought-to-grow-vegetables-idUSKBN1ZJ0XG.

Engindeniz, S. 2004. "The Economic Analysis of Growing Greenhouse Cucumber with Soilless Culture System: The Case of Turkey." *Journal of Sustainable Agriculture* 23 (3): 5–19.

Engindeniz, S. 2009. "Economic Analysis of Soilless and Soil-Based Greenhouse Cucumber Production in Turkey." *Scientia Agricola* 66 (5): 606–14.

Fukuda, N., M. Fujitan, Y. Ohta, S. Sase, S. Nishimura, and H. Ezura. 2008. "Directional Blue Light Irradiation Triggers Epidermal Cell Elongation of Abaxial Side Resulting in Inhibition of Leaf Epinasty in Geranium under Red Light Condition." *Scientia Horticulturae* 115 (2): 176–82.

Gabe, Vida. 2020. "Growth in the Time of COVID: How a Hydroponics Project Is Keeping Refugees in Chad Fed during the Pandemic." WFP Innovation Accelerator, December 18, 2020. https://wfpinnovation.medium.com/growth-in-the-time-of-covid-how-a-hydroponics-project-is-keeping-refugees-in-chad-fed-during-the-88d073f6ae0c.

Gichuhi, P. N., D. Mortley, E. Bromfield, and A. C. Bovell-Benjamin. 2009. "Nutritional, Physical, and Sensory Evaluation of Hydroponic Carrots (Daucus carota L.) from Different Nutrient Delivery Systems." *Journal of Food Science* 74 (9): 403–12.

Government of India. 2010. "Basic Animal Husbandry Statistics." Department of Animal Husbandry, Dairying and Fisheries, Ministry of Agriculture, Government of India, Krishi Bhawan, New Delhi.

Grafiadellis, I., K. Mattas, E. Maloupa, I. Tzouramani, and K. Galanopoulos. 2000. "An Economic Analysis of Soilless Culture in Gerbera Production." *HortScience* 35 (2): 300–03.

Graves, C. J. 1980. "The Nutrient Film Technique." *Horticultural Reviews* 5: 1–44.

Graves, C. J. 1983. "The Nutrient Film Technique." In *Horticultural Reviews*, edited by J. Janick, chapter 1. Avi Publishing Company, Inc. doi:10.1002/9781118060728.

Greer, L., and S. Diver. 2000. "Organic Greenhouse Vegetable Production: Appropriate Technology Transfer for Rural Areas." *Appropriate Technology Transfer for Rural Areas*, University of Arkansas, Fayetteville.

Gruda, N. 2009. "Does Soilless Culture Have an Influence on Product Quality of Vegetables?" *Journal of Applied Botany and Food Quality* 82: 141–47.

Heredia, N. A. 2014. "Design, Construction, and Evaluation of a Vertical Hydroponic Tower." California Polytechnic State University, San Luis Obispo.

Hoagland, D. R., and D. I. Arnon. 1950. "The Water Culture Method for Growing Plants without Soil." V.C347, College of Agriculture, University of California, Berkeley.

iGrow. 2020. "2020's Indoor Farm Venture Capital Bonanza." *iGrow News*, January 12, 2020. https://www.igrow.news/igrownews/2020s-indoor-farm-venture-capital-bonanza.

Intrado. 2021. "Hydroponics Market Size Worth $ 22.2 Billion by 2028; CAGR of 11.3%." Reports and Data, Intrado Global News Wire, March 3, 2021. https://www.globenewswire.com/en/news-release/2021/03/03/2186022/0/en/Hydroponics-Market-Size-Worth-22-2-Billion-By-2028-CAGR-of-11-3-Reports-and-Data.html.

Jensen, M. H. 1997. "Hydroponics." *HortScience* 32 (6): 1018–21.

Jensen, M. H. 1999. "Hydroponics Worldwide." *Acta Horticulturae* 481 (87): 719–29.

Jensen, M. H., and A. Malter. 1995. "Protected Agriculture: A Global Review." Technical Paper 253, World Bank, Washington, DC.

Jones Jr., J. B. 2016. *Hydroponics: A Practical Guide for the Soilless Grower*, second ed. Boca Raton, FL: CRC Press.

Kratky, B. A. 2009. "Three Non-Circulating Hydroponic Methods for Growing Lettuce." *Proceedings of the International Symposium on Soilless Culture and Hydroponics. Acta Horticulturae* 843: 65–72.

Ly, H. M. 2011. "Converting Soil Grown Production Methods to Hydroponics in Protected Cropping." No. 1014, Nuffield International, Taunton, Somerset, England.

Lykas, C., N. Katsoulas, P. Giaglaras, and C. Kittas. 2006. "Electrical Conductivity and pH Prediction in a Recirculated Nutrient Solution of a Greenhouse Soilless Rose Crop." *Journal of Plant Nutrition* 29: 1585–99.

MarketsandMarkets. 2019. "Hydroponics Markets Research Report." MarketsandMarkets Research, Northbrook, IL. https://www.marketsandmarkets.com/Market-Reports/hydroponic-market-94055021.html.

Mathias, M. C. 2014. "Emerging Hydroponics Industry." *Practical Hydroponics & Greenhouses* June: 18–21.

Ministry of Agriculture of Djibouti. 2019. "Results Report: Technical Assistance for Design and Implementation of Hydroponic Pilot Project in Djibouti." Ministry of Agriculture, Water, Fisheries and Livestock, Djibouti.

Morgan, J., R. R. Hunter, and R. O'Haire. 1992. "Limiting Factors in Hydroponic Barley Grass Production." In *Proceedings of the Eighth International Congress on Soilless Culture*, 241–61. Hunter's Rest, South Africa.

Mukhopad, Y. 1994. "Cultivating Green Forage and Vegetables in the Buryat Republic." *Mezhdunarodnyi Selskokhozyaistvennyi Zhurnal* 6: 51–52.

Naik, P. K., and N. P. Singh. 2013. "Hydroponics Fodder Production: An Alternative Technology for Sustainable Livestock Production against Impeding Climate Change." In *Compendium of Model Training Course Management Strategies for Sustainable Livestock Production against Impending Climate Change*, 70–75. November 18–25, Southern Regional Station, National Dairy Research Institute, Adugodi, Bengaluru, India.

Nederhoff, E., and C. Stanghellini. 2010. "Water Use Efficiency of Tomatoes in Greenhouses and Hydroponics." *Practical Hydroponics and Greenhouses* 115: 52–59.

Neff, M. M., C. Fankhauser, and J. Chory. 2000. "Light: An Indicator of Time and Place." *Genes & Development* 14: 257–71.

Pantanella, E., M. Cardarelli, G. Colla, and E. Marcucci. 2012. "Aquaponics vs. Hydroponics: Production and Quality of Lettuce Crop." *Acta Horticulturae* 927: 887–94. 10.17660/ActaHortic.2012.927.109.

Reddy, G. V. N., M. R. Reddy, and K. K. Reddy. 1988. "Nutrient Utilization by Milch Cattle Fed on Rations Containing Artificially Grown Fodder." *Indian Journal of Animal Nutrition* 5 (1): 19–22.

Resh, H. M. 1995. *Hydroponic Food Production: A Definitive Guidebook of Soilless Food-Growing Methods*, fifth ed. Santa Barbara, CA: Woodbridge.

Resh, H. M. 2013. *Hydroponic Food Production: A Definitive Guidebook of Soilless Food-Growing Methods*, seventh ed. Santa Barbara, CA: CRC Press.

Resh, H. M., and M. Howard. 2012. *Hydroponic Food Production: A Definitive Guidebook for the Advanced Home Gardener and the Commercial Hydroponic Grower*. Santa Barbara, CA: CRC Press.

Sanchez, S. V. 2007. "Evaluation of Curly Lettuce Cultivars Produced in Hydroponics Type NFT in Two Protected Environments in Ribeirão Preto (SP)." Master's thesis, Faculty of Agricultural and Veterinary Sciences, Universidade Estadual Paulista, São Paulo, Brazil. http://hdl.handle.net/11449/96944.

Selma, M. V., M. C. Luna, A. Martínez-Sánchez, J. A. Tudela, D. Beltrán, C. Baixauli, and M. I. Gil. 2012. "Sensory Quality, Bioactive Constituents and Microbiological Quality of Green and Red Fresh-Cut Lettuces (Lactuca sativa L.) Are Influenced by Soil and Soilless Agricultural Production Systems." *Postharvest Biology and Technology* 63 (1): 16–24.

Sgherri, C., S. Cecconami, C. Pinzino, F. Navari-Izzo, and R. Izzo. 2010. "Levels of Antioxidants and Nutraceuticals in Basil Grown in Hydroponics and Soil." *Food Chemistry* 123 (2): 416–22.

Shaw, N., D. Cantliffe, J. C. Rodriguez, and Z. Karchi. 2007. "Alternative Use of Pine Bark Media for Hydroponic Production of Galia Muskmelon Results in Profitable Returns." In *Third International Symposium on Cucurbits, Acta Horticulturae*, 259–65.

Sneath, R., and F. Mclntosh. 2003. "Review of Hydroponic Fodder Production for Beef Cattle." Queensland Government Department of Primary Industries, Dalby, Queensland, Australia.

Spensley, K., G. W. Winsor, and A. J. Cooper. 1978. "Nutrient Film Technique: Crop Culture in Flowing Nutrient Solution." *Outlook on Agriculture* 9 (6): 299–305.

Treftz, C., and S. T. Omaye. 2016. "Comparison between Hydroponic and Soil Systems for Growing Strawberries in a Greenhouse." *International Journal of Agricultural Extension* 3 (3): 195–200.

Verner, D. 2016. "Could a Livelihood in Agriculture Be a Way for Refugees to Move from Surviving to Thriving?" *Arab Voices* (blog), May 23, 2016. https://blogs.worldbank.org/arabvoices/livelihood-in-agriculture-refugees.

Verner, D., S. Vellani, A.-L. Klausen, and E. Tebaldi. 2017. "Frontier Agriculture for Improving Refugee Livelihoods: Unleashing Climate-Smart and Water-Saving Agriculture Technologies in MENA." World Bank, Washington, DC. https://elibrary.worldbank.org/doi/abs/10.1596/29753.

Welthungerhilfre. 2020. "Simple but Innovative Hydroponic Gardens in Sudan." Welthungerhilfre Blog, January 16, 2020. https://www.welthungerhilfe.org/news/latest-articles/2020/simple-but-innovative-hydroponic-gardens-in-sudan/.

WFP Kenya (World Food Programme Kenya). 2020. "Hydroponics Pilot Project Report." WFP Kenya, Nairobi.

Wohlgenant, M. K. 2001. "Marketing Margins: Empirical Analysis." In *Handbook of Agricultural Economics*, vol. 1, part B, edited by B. L. Gardner and G. C. Rausser, 933–70. Elsevier.

Wootton-Beard, P. 2019. "Growing without Soil: An Overview of Hydroponics." Farming Connect, Business Wales.

World Bank Group. 2019. *World Bank Group Strategy for Fragility, Conflict, and Violence 2020–2025*. Washington, DC: World Bank Group.

CHAPTER SIX

Ways Forward

To increase food security in Africa, particularly in food insecure countries affected by fragility, conflict, and violence (FCV), calls for sustainably producing nutritious food and increasing access for all. The circular food economy is a model for achieving this within planetary boundaries and promoting green, resilient, and inclusive development, known as "GRID." The GRID framework enables social and economic transformations that can reduce poverty and food insecurity and promote equality, resilience, and shared prosperity while leveraging private sector support. The process of implementing a circular food economy based on frontier agricultural technologies is extensive. However, there are two distinct phases in achieving that outcome. The first phase is to establish and pilot the system. Establishing the necessary foundations of institutions and frameworks will carry the effort forward, while piloting the frontier agricultural systems will remove the inefficiencies and demonstrate and enhance the benefits. The second phase is to scale up frontier agricultural production systems at large enough levels to shift existing linear food economies into circular food economies. These two phases would address the major factors that constrain the widespread adoption of insect and hydroponic farming in Africa. Figure 6.1 shows how the two phases—(1) establishing and piloting and (2) scaling—propel the circular food economy.

Currently, various factors constrain widespread adoption of insect and hydroponic farming in Africa. These factors create barriers to entry that are specific to a farmer's location and access to markets and technologies. Many of these factors are a result of the relative newness of both technologies. In Africa, foraging for and consumption of foraged plants and insects are not new, but insect and hydroponic farming are still nascent industries. For example, according to the

FIGURE 6.1 Developing a Circular Food Economy

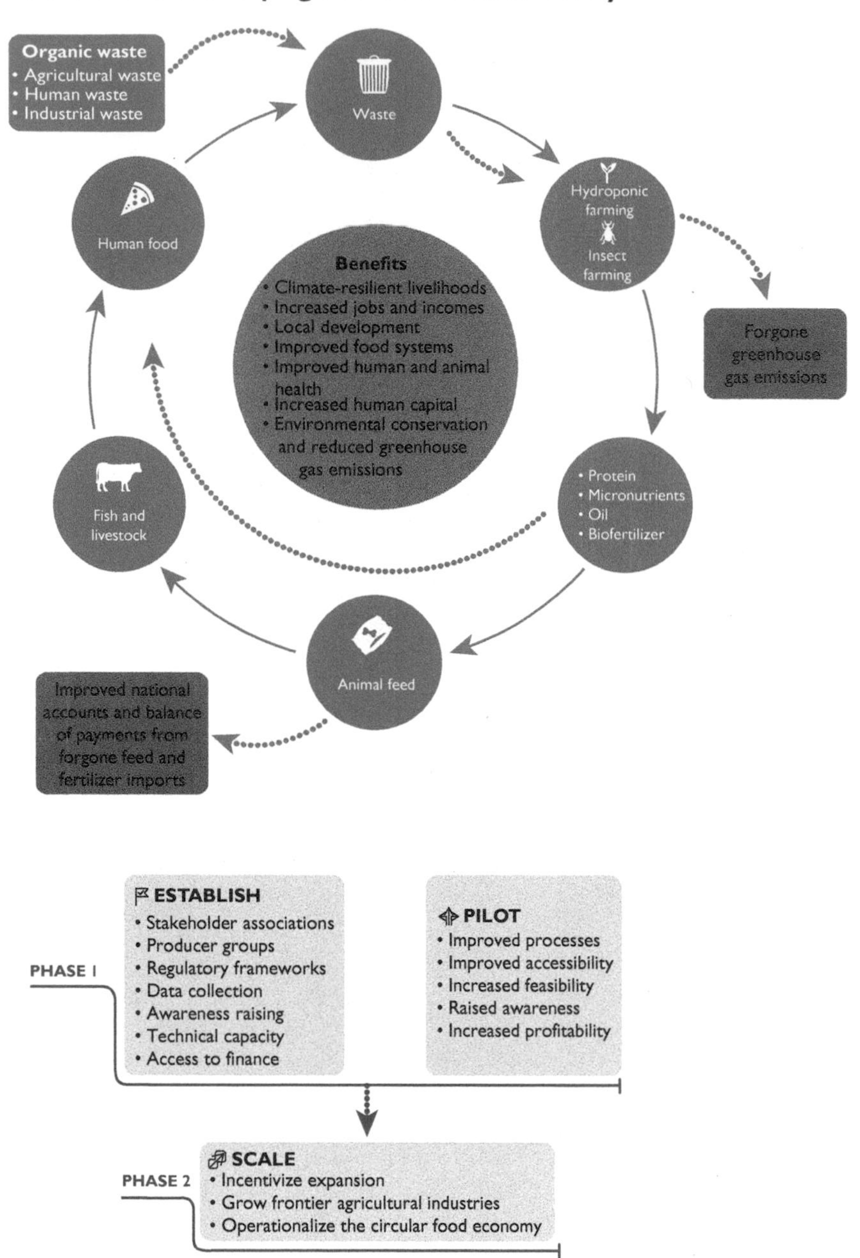

Source: Original figure for this publication.

surveys that were carried out for this report, farming crickets has been practiced in Africa for only 20 years. Farming other insects, such as black soldier flies, is even more recent. This newness means that the industry is still underdeveloped and not well understood by many. This contributes to seven main constraints to the widespread adoption of these technologies in Africa.

1. There is a general lack of knowledge and learning systems. The farm-level surveys show that African farmers are willing to farm insects but lack the knowledge on how to go about it, including knowledge on the production cycle and gains in nutrition for humans and animals. The same is true for hydroponics. There are also knowledge gaps for both technologies in technical know-how and market comprehension and few learning or monitoring and evaluation systems.
2. There are few organizations and institutions to transfer this knowledge. This includes the absence of formal training opportunities and informal peer-to-peer learning through producer groups. However, the surveys show that knowledge and learning have increased in areas located near insect research facilities in Ghana and Kenya.
3. Regulatory frameworks on food safety and production methods are weak for both the insect farming and hydroponic farming sectors in Africa. This makes it difficult for the industry to develop or for potential farmers to enter the market because there is no clear guidance on production or food safety protocols. Other parts of the world have tried to correct this. For example, in Thailand, the government developed a Good Agriculture Practice certification for farmers. Moreover, different insect species require different production processes; therefore, regulations must guide farmers on these processes to produce insects that can be safely consumed.
4. There are no notable champions among African governments supporting these sectors or guiding potential stakeholders. For example, as of mid-2021, there were no insect or hydroponic sector plans or strategies among African governments. One of the problems is that insect and hydroponic farming are still emerging sectors and have not yet caught the attention of many policy makers. The government of the Republic of Korea is one of the few that has developed an insect sector plan. Such a plan could serve as a model for African governments to develop supply chains and manage risks.
5. As described in chapter 1 of this report, farming subsidies still artificially prop up conventional farming. These subsidies lower the costs of conventional farming at the expense of the environment and frontier agricultural technologies. By contrast, governments in some higher-income countries—such as Denmark, Korea, and the Netherlands—provide funds to kick-start insect farming.
6. There is insufficient access to finance and, as described in chapters 3 and 5, there are very real cost constraints for farmers to access inputs or expand operations. The farmer-level surveys show that many farmers would like to expand their production, but they cannot do so because of a lack of finance.

The surveys also show that high start-up costs and limited access to inputs—such as eggs, substrates, or hardware—can set back production. For example, in Madagascar, farmers who were interviewed said they had to travel more than 50 kilometers to buy insect eggs. In Kenya, the research team observed that many of these input and scaling constraints were removed when farmers had greater access to finance.

7. There is still some cultural aversion toward consuming insects, although this seems to be more of an issue in Europe and North America than in Africa. The survey results show that most of the interviewed farmers in Africa were comfortable with the idea of eating insects or feeding them to livestock. In Zimbabwe, 90 percent of people eat insects on a regular basis (Dube et al. 2013).

PHASE 1: ESTABLISHING AND PILOTING

To establish frontier agricultural technologies—specifically, insect farming and crop hydroponics—requires several key actions. These include, but are not limited to, (1) forming producer groups; (2) building technical capacity among producers; (3) providing access to finance; (4) forming entomophagy and hydroponic associations; (5) raising public awareness of the social, economic, and environmental benefits of frontier agriculture; (6) strengthening regulatory frameworks; (7) monitoring and evaluating; and (8) piloting programs to increase the functionality, accessibility, and affordability of frontier agricultural production. These actions are described in the following subsections.

Forming Community-Level Producer Groups

National associations can assist small-scale producers in forming community-level producer groups. These groups could register as legal entities that supply frontier agricultural products to local communities and regional markets. The producer groups could provide several services to local producers, including trainings, information sharing, and group savings programs. Most important, the producer groups can advocate for the needs of small-scale producers to the national association and eventually to the government on policy.

Building Technical Capacity on Frontier Agricultural Technologies among Producers

Currently, insect farming and plant hydroponics are not well known or widely practiced in Africa. As a result, local and international stakeholders must disseminate information through collaboration, information sharing, and extension services. As this report has shown, many African farmers are already farming insects or have expressed an interest and willingness to try—but still lack the knowledge to expand or get started. Technical knowledge training would fill that gap. Instruction would include how to set up the frontier technology's infrastructure, how to tend the system, how to process and add value to products, how to market and sell products in the market, and how to carry out these practices safely, hygienically, and sustainably. Many universities in Africa—from Accra to Nairobi and from Bujumbura to Chinhoyi—already have capacity in these areas. This outreach training would likely also inform farmers of relevant policies and regulations. To accomplish this, the technical knowledge must fall within a framework of long-term partnerships and training transfers with research and technical institutions. These partnerships would eventually play a key role in piloting insect and hydroponic farms.

Providing Farmers Access to Finance

Beyond technical knowledge, rural farmers will likely need financing to start operations, even for the simplest systems, which all require tools and input materials. Initially, farmers may require grants or low-interest loans, but eventually, once frontier agriculture shows returns, risk guarantees can be used. With the support of financial institutions and international donors, governments can embed financial systems in existing rural, social, agricultural, and private sector development programs.

Forming National Entomophagy and Hydroponic Associations

The first step to developing a country's frontier agricultural technologies is bringing together existing stakeholders. These stakeholders should include representation from the government, the private sector, and civil society. National entomophagy and hydroponic associations would organize stakeholders around the general mandate of mainstreaming frontier agricultural technologies in a given country. This would be an important first step in African FCV-affected countries where there is a lack of guidance and limited regulatory frameworks for selling and consuming frontier agricultural products, especially farmed insect products. The associations could develop ethics and food safety guidelines for each industry, respond to emerging issues, and generally lay the foundation for mainstreaming frontier agricultural technologies in a circular food economy. The associations would provide producers access to information on best practices for hydroponics, rearing and processing farmed insects, sustainably managing resources, utilizing new technologies and equipment, sourcing and pricing inputs, acquiring pertinent market data on local and regional demand, and developing online marketplaces for suppliers and producers. The associations would provide a forum for stakeholders to coordinate production, processing, and sales to improve supply chain efficiency and competitiveness. The associations would also provide advocacy services to promote certain sector strategies and national laws and regulations. This advocacy would be strengthened through awareness-raising campaigns and market links with wholesale and retail supply chains.

Implementing a Communications Strategy to Raise Public Awareness on the Social, Economic, Climatic, and Environmental Benefits of Frontier Agriculture

The stakeholder associations should lead awareness-raising efforts in African countries to inform the public on the benefits of circular economy and the insect farming and plant hydroponics sectors. Such awareness raising is crucial for farming insects for human food or animal feed since there still may be some cultural aversion or common misperception that insects are unfit for human consumption. The associations would work with governments, civil society organizations, and international donors to raise awareness and challenge these biases. These efforts should leverage forward-looking and progressive approaches to awareness raising, such as Social and Behavior Change Communication Theory or others. History demonstrates that it is possible to change people's opinions about foods that were once considered undesirable. For example, in the United States, at one time, people in Maine, long known for its lobster industry, did not consume much of the lobster themselves, but when the railway came they managed to export it to other states. It was only during the 1880s that lobster shed some of its negative reputation as upscale diners in Boston and New York began to pay large sums for it (History.com 2018). A similar dynamic could happen with farmed insects. Cultural influencers, such as chefs, could help raise the status of insects as a food that is suitable and even desirable for humans. Some chefs are already doing this, for example at Restaurant Noma in Denmark, Yoosung Hotel in Korea, and the Victoria Falls Hotel in Zimbabwe. The survey results show that Africans are less likely than people from Western countries to hold negative perceptions of insect consumption.

Strengthening the Regulatory Frameworks for Relevant Industries

Frontier agricultural technologies may require government regulations for the industries' markets, production, food safety, and environmental protection. As of 2021, such regulations do not exist in Africa or are limited in what agricultural products they cover. As such, there are very few models to follow. For example, the European Union only recently implemented novel food legislation covering insect food products. Regulatory frameworks also govern the many stakeholders involved in frontier agriculture. This includes members of national associations and producer groups, but also consumers, investors, wholesalers and retailers, feed producers, food producers, research institutions, entrepreneurs, policy makers, government agencies, individual farmers, food service workers, civil society organizations, and the international community, among others. Figure 6.2 shows what a potential institutional and regulatory framework might look like for the farmed insect food and feed industry. With the broad mandate to regulate frontier agriculture, the framework should also ensure policy coherence; establish an incentive structure to grow insect

FIGURE 6.2 Institutional and Regulatory Framework for Farmed Insects as Food and Feed

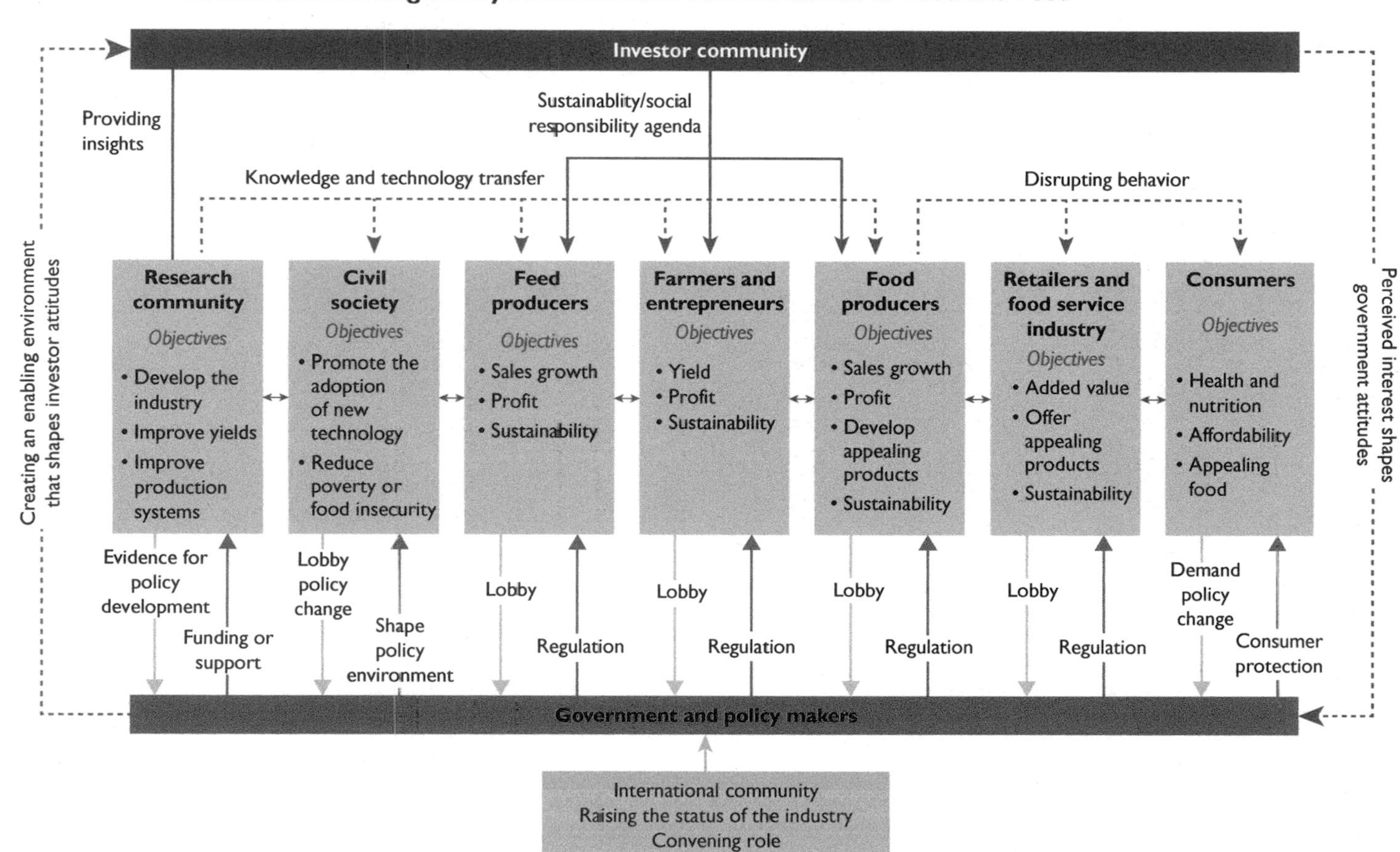

Source: Original figure for this publication.

farming and hydroponic agriculture; produce guidelines for insect rearing, handling, harvesting, and processing; establish criteria for introducing new insect or hydroponic products into the food supply chain; define an environmental conservation strategy for both industries; and establish sanitary and hygiene requirements for food products and substrates, including metal tracing, especially in informal markets. Likewise, there may be room to adjust or reverse farm subsidies for conventional agriculture when those laws negatively affect the environment and stunt circular food industries.

Monitoring and Evaluating All Aspects of Frontier Agricultural Technologies

As the research team discovered while compiling this report, data on insect and hydroponic farming are scarce, compared with data on conventional forms of agriculture. Occasionally, studies emerge on the science of both technologies, but there is very little practical information on markets, value chains, or the general business environment. This is especially true for individual countries in Africa, where there is already a dearth of data on many subjects, let alone largely unknown, underutilized subjects like frontier agriculture. As a result, individual countries, led by the respective government and national stakeholder associations, should invest in data collection and other monitoring and evaluation efforts. This will inform the government's decision making and producers' and sellers' business development.

Establishing Pilot Programs to Increase the Functionality, Accessibility, and Affordability of Frontier Agricultural Production

There are still some uncertainties surrounding the feasibility of frontier agricultural technology in Africa. Although the benefits are clear, some uncertainties lie in the implementation. Piloting will provide an opportunity to increase knowledge on frontier agricultural operations, improve the industries' functionality, and make production systems more accessible and cost-effective. Pilots will also demonstrate the successes of frontier agricultural systems, thus raising awareness of the systems' benefits in the process. The pilot projects will help define the following aspects of farmed insect and hydroponic food systems:

- Understanding the accessibility of inputs for constructing and maintaining farming systems
- Demonstrating the residue and organic waste recovery potential from the production phase of both systems
- Identifying alternative substrate sources for insect farms, such as brewers' spent grains, green market wastes, food processing wastes, and other local sources of organic waste
- Determining the ideal climates and environmental conditions for insect and hydroponic plant production

- Establishing systems to monitor and map waste streams that would be used as inputs in frontier agricultural systems
- Determining the most effective entry points for insect and hydroponic farming into existing industries, such as waste management, animal feed production, human food production, and others
- Establishing effective means of communicating among farmers, processers, and consumers
- Creating networks among farmers to provide a stable supply of food products to the market
- Improving pest and disease management in both systems
- Refining processes for detecting harmful trace metals in foods and substrates
- Pursuing energy efficient ways to carry out production, including the use of renewable energy
- Creating a digital portal to match supply and demand for inputs, products, and substrates

The insect farming project in Kenya's Kakuma refugee camp could serve as a model for other insect production pilots (see box 4.1). The pilot trained refugees in cricket farming techniques. The project started out with a pilot cricket farm and provided training to 15 refugee household heads. Since then, the project has trained more than 80 household heads in rearing and processing farmed insects. These household heads—who have fled from conflict in countries like Burundi, the Democratic Republic of Congo, and South Sudan—are now producing crickets for household consumption and animal feed. DanChurchAid is planning to scale up the initiative by training more farmers and distributing starter kits to more refugee households. The project shows the potential for insect farming to provide livelihoods and incomes for marginalized communities, even in FCV situations.

The hydroponic project in West Bank and Gaza could serve as a model for other hydroponic pilots (see box 5.2).[1] In 2012, the pilot established nutrient film technique and wicking bed production systems to increase local incomes, nutrition, food security, women's empowerment, and the competitiveness of the agricultural cooperatives sector. The pilot established 35 nutrient film technique units and 52 wicking bed units with marginalized and underprivileged families in remote areas of the Bethlehem and Hebron governorates. The pilot included education modules at a local technical school to train students in these technologies. The families consumed most of the food that was produced and sold the surplus to local markets. This pilot has since advanced and produces different crops in new systems.

PHASE 2: SCALING

Once the pilots are completed and the lessons learned from those efforts are recorded, they can be applied to launch larger-scale frontier agricultural operations.

Planners can choose among the many existing farming models when scaling up and expanding insect and hydroponic farming. These models include a large-scale insect farming model where larger individual operations produce and process insect products; an organizational model where small farmers organize into a larger association that aggregates outputs, processes those outputs collectively, and sells them wholesale; a contract farming model where wholesalers and processors contract farmers to produce insects; and many other models. Regardless of the model, planners could assist and encourage farmers, both new and existing, in entering or expanding their frontier agricultural operations to reach maximum potential within a circular food economy. Throughout the scaling process, it is important to monitor and evaluate the results and record the lessons learned to maximize the industries' outcomes and improve future expansion efforts.

Equity should be ensured when scaling up frontier agricultural operations. The larger these operations become relative to conventional fish, livestock, and agricultural practices, the more they provide jobs and incomes, reduce greenhouse gas emissions and environmental degradation, and save hard currency that was previously reserved for importing fertilizers and soybeans for animal feed. At the same time, inequalities could be exacerbated as the industry grows. Therefore, a key step is to integrate small producers into frontier agricultural value chains as these markets develop. The public sector could play a key role in this integration by investing in research and development, training, and regulatory frameworks and the frontier agriculture sector, much as the Korean government has done to develop the country's insect farming sector. Once the scaling process includes proper measures to ensure equity, the regulatory framework could include incentives for establishing new, larger-scale operations. Qualifying for incentives could be predicated on the likelihood of the producer achieving high volumes, employment, cost efficiency, and market demand. It is in this context that Africa has huge potential for scaling frontier agricultural industries and emerging as a global leader in insect-based protein farming and processing.

NOTE

1. Applied Research Institute–Jerusalem, personal communication, 2021.

REFERENCES

Dube, S., N. R. Diamini, A. Mafung, M. Mukai, and Z. Dhlamini. 2013 "A Survey on Entomophagy Prevalence in Zimbabwe." *African Journal of Food, Agriculture, Nutrition and Development* 13 (1): 7242–53.

History.com. 2018. "A Taste of Lobster History." History.com, August 22, 2018. https://www.history.com/news/a-taste-of-lobster-history.

ECO-AUDIT
Environmental Benefits Statement

The World Bank Group is committed to reducing its environmental footprint. In support of this commitment, we leverage electronic publishing options and print-on-demand technology, which is located in regional hubs worldwide. Together, these initiatives enable print runs to be lowered and shipping distances decreased, resulting in reduced paper consumption, chemical use, greenhouse gas emissions, and waste.

We follow the recommended standards for paper use set by the Green Press Initiative. The majority of our books are printed on Forest Stewardship Council (FSC)–certified paper, with nearly all containing 50–100 percent recycled content. The recycled fiber in our book paper is either unbleached or bleached using totally chlorine-free (TCF), processed chlorine–free (PCF), or enhanced elemental chlorine–free (EECF) processes.

More information about the Bank's environmental philosophy can be found at http://www.worldbank.org/corporateresponsibility.